THE ART OF WATCH REPAIR

INCLUDING DESCRIPTIONS OF THE WATCH MOVEMENT, PARTS OF THE WATCH AND COMMON STOPPAGES OF WRIST WATCHES

British Library Cataloguing-in-Publication Data
A catalogue record for this book is available from the
British Library

A History of Clocks and Watches

Horology (from the Latin, Horologium) is the science of measuring time. Clocks, watches, clockwork, sundials, clepsydras, timers, time recorders, marine chronometers and atomic clocks are all examples of instruments used to measure time. In current usage, horology refers mainly to the study of mechanical time-keeping devices, whilst chronometry more broadly included electronic devices that have largely supplanted mechanical clocks for accuracy and precision in time-keeping. Horology itself has an incredibly long history and there are many museums and several specialised libraries devoted to the subject. Perhaps the most famous is the *Royal Greenwich Observatory,* also the source of the Prime Meridian (longitude 0° 0' 0"), and the home of the first marine timekeepers accurate enough to determine longitude.

The word 'clock' is derived from the Celtic words *clagan* and *clocca* meaning 'bell'. A silent instrument missing such a mechanism has traditionally been known as a timepiece, although today the words have become interchangeable. The clock is one of the oldest human interventions, meeting the need to consistently measure intervals of time shorter than the natural units: the day, the lunar month and the year. The current sexagesimal system of time measurement dates to approximately 2000 BC in Sumer. The Ancient Egyptians divided the day into two twelve-hour periods and used large obelisks to track the movement of the sun. They also developed water clocks, which had also been employed frequently by the Ancient Greeks, who called them 'clepsydrae'. The Shang Dynasty is also believed to have used the outflow water clock around the same time.

The first mechanical clocks, employing the verge escapement mechanism (the mechanism that controls the rate of a clock by advancing the gear train at regular intervals or 'ticks') with a foliot or balance wheel timekeeper (a weighted wheel that rotates back and forth, being returned toward its centre position by a spiral), were invented in Europe at around the start of the fourteenth century. They became the standard timekeeping device until the pendulum clock was invented in 1656. This remained the most accurate timekeeper until the 1930s, when quartz oscillators (where the mechanical resonance of a vibrating crystal is used to create an electrical signal with a very precise frequency) were invented, followed by atomic clocks after World War Two. Although initially limited to laboratories, the development of microelectronics in the 1960s made quartz clocks both compact and cheap to produce, and by the 1980s they became the world's dominant timekeeping technology in both clocks and wristwatches.

The concept of the wristwatch goes back to the production of the very earliest watches in the sixteenth century. Elizabeth I of England received a wristwatch from Robert Dudley in 1571, described as an arm watch. From the beginning, they were almost exclusively worn by women, while men used pocket-watches up until the early twentieth century. This was not just a matter of fashion or prejudice; watches of the time were notoriously prone to fouling from exposure to the elements, and could only reliably be kept safe from harm if carried securely in the pocket. Wristwatches were first worn by military men towards the end of the nineteenth century, when the importance of synchronizing manoeuvres during war without potentially revealing the plan to the enemy through signalling was increasingly recognized. It was clear that using pocket watches while in the heat of battle or while mounted on a horse was impractical, so officers began to strap the watches to their wrist.

The company H. Williamson Ltd., based in Coventry, England, was one of the first to capitalize on this opportunity. During the company's 1916 AGM it was noted that '...the public is buying the practical things of life. Nobody can truthfully contend that the watch is a luxury. It is said that one soldier in every four wears a wristlet watch, and the other three mean to get one as soon as they can.' By the end of the War, almost all enlisted men wore a wristwatch, and after they were demobilized, the fashion soon caught on - the British *Horological Journal* wrote in 1917 that '...the wristlet watch was little used by the sterner sex before the war, but now is seen on the wrist of nearly every man in uniform and of many men in civilian attire.' Within a decade, sales of wristwatches had outstripped those of pocket watches.

Now that clocks and watches had become 'common objects' there was a massively increased demand on clockmakers for maintenance and repair. Julien Le Roy, a clockmaker of Versailles, invented a face that could be opened to view the inside clockwork – a development which many subsequent artisans copied. He also invented special repeating mechanisms to improve the precision of clocks and supervised over 3,500 watches. The more complicated the device however, the more often it needed repairing. Today, since almost all clocks are now factory-made, most modern clockmakers *only* repair clocks. They are frequently employed by jewellers, antique shops or places devoted strictly to repairing clocks and watches.

The clockmakers of the present must be able to read blueprints and instructions for numerous types of clocks and time pieces that vary from antique clocks to modern time pieces in order to fix and make clocks or watches. The trade requires fine motor coordination as clockmakers must frequently work on devices with small gears and fine machinery, as well as an appreciation for the original art form. As is evident from this very short history of clocks and watches, over the centuries the items themselves have changed – almost out of recognition, but the importance of time-keeping has not. It is an area which provides a constant source of fascination and scientific discovery, still very much evolving today. We hope the reader enjoys this book.

DESCRIPTION OF A WATCH MOVEMENT.

It is not proposed in this book to enter into the question of the history of the watch, nor to discuss who invented this or that portion of it, but simply to take the modern watch as it is, and describe as clearly as possible how it works and how to repair and keep it in order.

It will, perhaps, be well first to describe, in general terms, the mechanism of a watch, and for this purpose a Geneva "bar" movement will be used as an illustration. Fig. 1 shows such a movement. The term "movement," it may be explained, is applied to the *works* of a watch as distinguished from the case.

This particular movement is chosen, as its "bar" construction enables all the wheelwork to be seen. The mechanism of this movement may be divided into four portions. *First*, the motive power; *second*, a train of wheels to transmit the power; *third*, an escapement and balance to control the power; and *fourth*, motion work and hands to record the revolutions of the train wheels upon the dial.

FIG. 1.—GENEVA BAR MOVEMENT.

The Motive Power.—This, in all watches, is a mainspring. A mainspring is a thin and flat strip of steel, hardened and tempered to give the maximum of strength and elasticity. It is coiled up around a steel centre arbor, to which its eye is hooked, and enclosed in a box or "barrel," to the inside of which its outer end is attached. If a barrel, containing such a spring, be held firmly. while the centre arbor is turned round, coiling up the spring tightly around it, until the outer end pulls hard at its attachment, as at A (Fig. 2), the barrel, when released, will revolve in the direction that the spring pulls it, until the spring has unwound itself and is prevented by the containing barrel from unwinding further, as at B (Fig. 2). The number of complete revolutions thus made by a watch barrel with an average mainspring is five, and the number of revolutions used in driving the watch for twenty-four hours is generally three, thus leaving two to spare.

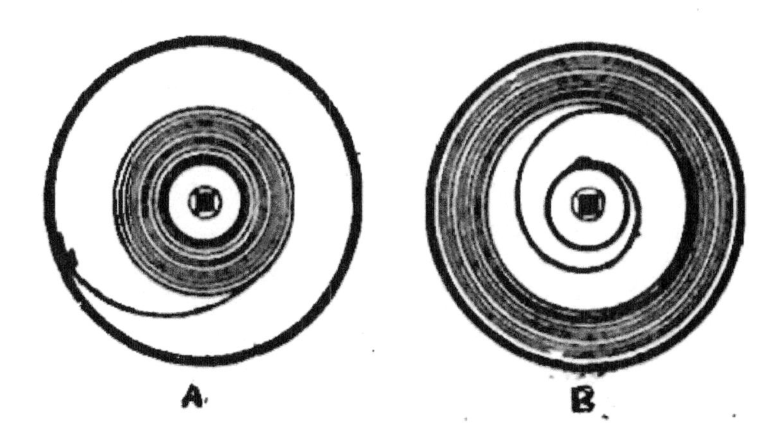

FIG. 2.—MAINSPRING IN ITS BARREL.

There are three principal methods of making a mainspring drive a watch. The first method, and the one adopted in the movement illustrated in Fig. 1, is to make the barrel into a toothed wheel by cutting teeth around its circumference. The barrel then becomes the first or "great wheel" of the watch train. This is termed a "going barrel." In a watch with this arrangement the barrel arbor is squared, and to wind the watch a key is placed upon it, and it is turned round three or four complete revolutions, being held by "clickwork." During the going of the watch, the barrel arbor is stationary and the barrel turns round, hence the term "going-barrel." "Clickwork" is the name given by watchmakers to an arrangement of a ratchet and pawl, the latter, in watches, being termed a "click." Fig. 3 shows a clickwork arrangement. In the figure A is the ratchet, B the click, and C the click-spring. In many watches the click and spring are in one piece, as in Fig. 1, but the action remains the same. In the second method the barrel is stationary while the arbor revolves, carrying with it a separate toothed wheel. In some watches on this plan the barrel is turned round in the act of winding the watch, and in others it is merely a sink recessed out in the solid watch plate, and, of course, a fixture. This method of driving is used mainly in American watches. The third method is for the barrel to be merely a drum, driving the watch by means of a chain wound upon it. This indirect and somewhat unsatisfactory method was adopted in order to equalize the force or pull of the spring. When a mainspring is fully wound up it exerts its maximum force. As it unwinds the force becomes gradually less and less, until it is zero. In the old watches the force of the mainspring directly affected the timekeeping of the watch, hence it was necessary to introduce some arrangement to equalize it. The arrangement adopted is shown in Fig. 4. A is the barrel containing the mainspring, B is the chain, C is the "fusee." The fusee is cone-shaped pulley having a continuous spiral groove cut upon it. The chain runs in this groove. When the spring is wound up the chain is on the fusee and pulls at its smallest diametei, thus exerting but a small leverage upon the fusee, to which is attached the first or main wheel of the watch train. Fig. 4 shows the arrangement when wound up. As it unwinds the barrel revolves and unwinds the chain from the fusee, coiling it up on itself. During this process the chain gets lower and lower upon the fusee body until, when nearly run down, it pulls upon the largest diameter of the cone, thus giving the diminished force of the mainspring an advantage in leverage. If the proportions of the cone are suited to the mainspring, it is possible by this means to have a constant force driving the watch throughout the twenty-four hours. With verge watches the fusee was a necessity. Though foreign makers quickly found that with all other kinds of watches the fusee was unnecessary, English makers, almost without exception, continued using it with lever watches for many years, and some use it now. In marine chronometers it is still in use.

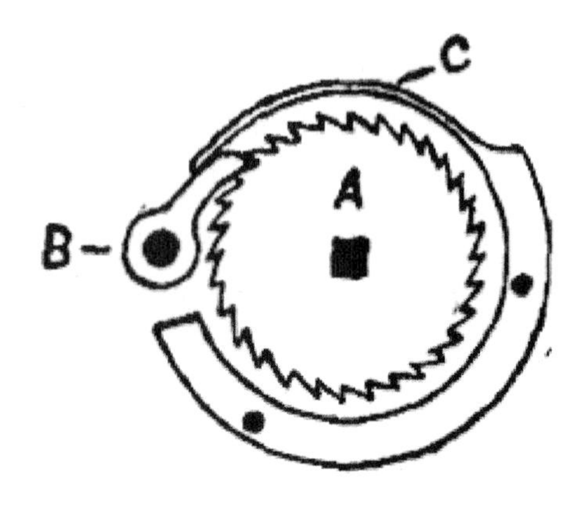

FIG. 3.—WINDING CLICKWORK.

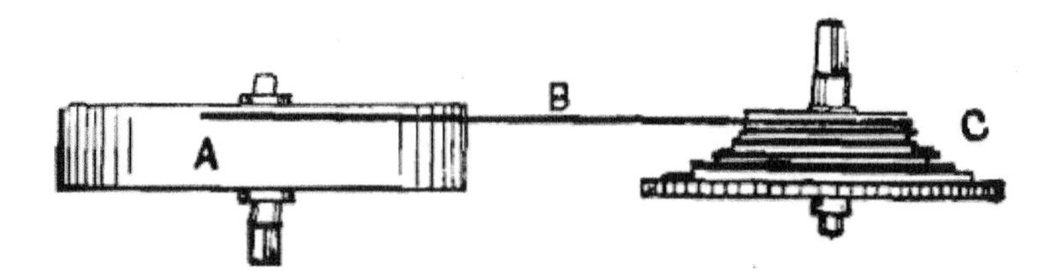

FIG. 4.—BARREL AND FUSEE.

The Train.—The mainspring thus, either directly or indirectly, drives the main wheel, which is the first wheel of the watch train. This wheel, in an average watch, turns once in eight hours. It gears into the centre pinion of the watch, causing the latter to revolve eight times to once of the main wheel, and thus turn once in one hour. This is effected by the main wheel having eight times as many teeth as the centre pinion has leaves. The centre pinion, as its name implies, occupies the centre of the watch, and its axis or "arbor" projects through the dial, and has the minute hand affixed to it. Upon the same arbor, with the centre pinion, is the centre wheel, the second wheel of the train, centre wheel and pinion forming one and revolving together. In the same way, the centre wheel drives the third wheel and pinion, and causes the latter to revolve eight times in one hour, or one revolution in seven and a half minutes, the centre wheel having eight times as many teeth as the third pinion has leaves. The third wheel, again, drives the fourth wheel, and has seven and a half times as many teeth as the fourth pinion has leaves. The fourth wheel and pinion therefore perform one revolution in one minute. A prolongation of one pivot of the fourth pinion projects through the dial and carries the seconds hand. The fourth wheel, in its turn, drives the scape pinion and wheel, causing the latter to perform ten revolutions in one minute. This completes the watch train and brings us to the escapement. In Fig. 1 all these wheels are visible, and the above explanation can be followed by reference to them. The wheels in most watches are of hard brass; a few have German silver or nickel wheels, and a few have wheels of a special alloy combining lightness and strength, such as aluminiumbronze. The pinions which they drive are of fine quality steel, hardened and tempered. The axis of a wheel is, in watchwork, called its "arbor." The pinions of the train wheels are in one piece with their arbors. Upon the ends of the arbors fine pivots are turned. These run either in pivot holes drilled in the plates or bars, or else in jewel holes, to diminish friction and reduce wear. A jewel hole is a small circular plate of garnet, sapphire, or ruby let into the brass of the watch frame. It is perforated in its centre with a fine, true, and polished hole, in which the pivot runs. Fig. 5 shows a wheel and pinion running in jewel holes. A is the wheel, B the pinion, C C the jewel holes.

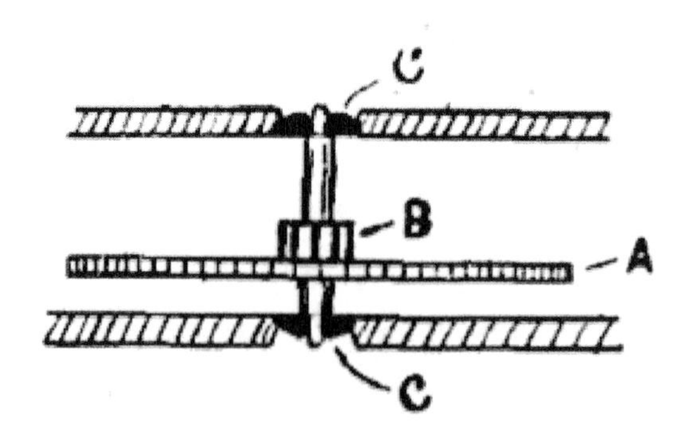

FIG. 5.—WHEEL AND PINION.

The teeth of the wheels and the leaves of the pinions are cut to very exact curves, so as to ensure a smooth and even motion when they are running together.

The Escapement and Balance.—It is obvious that, given a mainspring and a train of wheels such as that just described, if the mainspring were wound up the train would run at full speed, the spring unwinding itself in a few moments. Some arrangement is therefore necessary to check it. In a watch this checking mechanism is termed an "escapement," the duty of the escapement being to allow only one tooth of the scape wheel to pass at a time, and that at perfectly regular intervals.

The duty of measuring and regulating the intervals is performed by the balance and hairspring. The balance is a flywheel, mounted upon an axis having extremely fine pivots running in jewel holes. It is controlled by a hairspring. The hairspring is a flat spiral of thin steel wire. Its inner end is affixed to a collet upon the axis of the balance. Its outer end is fixed to a stud rigidly fastened to some part of the watch frame. If a balance, mounted and fitted in this way, be given a turn round in one direction and then let go, it will return under the influence of the hairspring and go nearly as far again in the reverse direction until its force is spent. The spring, then, causes it to return again, and it will be kept vibrating for some time before it finally comes to rest. It, in fact, acts in much the same way as a pendulum, which, when set swinging, continues to swing, traversing a smaller and smaller arc each time, until it is brought to rest by friction at its point of suspension and the resistance of the air.

The short intervals of time (generally one-fifth of a second) thus measured by the balance, and its spring are always very nearly equal, and, under some conditions, exactly equal, whatever the distance traversed by the balance may be.

The escapement divides up the power of the mainspring into small portions, and delivers these portions to the balance at each of its vibrations, giving it, as it were, a little helping push—an "impulse"—as it comes round each time.

Thus, through the medium of the escapement, the mainspring keeps the balance vibrating, and the balance regulates the running of the train.

The balance and hairspring are well seen in Fig. 1.

The Motion Work.—All the mechanism before described would be of little practical use unless the revolutions of the various wheels could be recorded in some way. The centre arbor, it has been seen, revolves once in one hour. Therefore a hand affixed to it will travel round the dial and serve to show the minutes. By gearing down from this arbor with a pair of wheels and pinions whose combined ratios are as 1 to 12, one pair being usually 1 to 4 and the other pair 1 to 3, and placing another hand upon the arbor of the last wheel, the hours 1 to 12 can be also shown in the usual way. These reducing wheels are termed the "motion work," and are not visible in Fig. 1, being hidden between the dial and the watch plate.

This is a brief description of a simple form of watch movement, and it will easily be believed that a great many different materials are used in its construction and in the processes of repairing it; also that it takes but a very little to upset its action or stop it altogether. Merely to enumerate all the possible faults and inaccuracies to which a watch is subject would fill many pages, and it is the purpose of this book to describe them all in detail and the way to overcome them.

THE PARTS OF A WATCH

With high-class watches, one can expect a very close rate of time under both extreme and normal conditions, but the inexpensive watch, by reason of its condition, cannot be expected to keep time within several seconds a day. A keen student, however, will soon be able to classify the various grades and execute work accordingly. Any new parts should be faithfully copied, if it is not possible to obtain standard material, in order to maintain the standard of the watch. Good work always reflects credit on the repairer.

FIG. 43.—A MODERN SWISS LEVER MOVEMENT OF THE POPULAR 10 1/2 LIGNE SIZE (SHOWN GREATLY ENLARGED).

Special Names for Parts.—The first step is to become thoroughly acquainted with the numerous components of an ordinary watch. Many parts have special names, and to be conversant with them will often save considerable time when ordering new material. The enlarged illustration on the opposite page depicts a modern Swiss Lever movement of the popular 10 1/2-ligne size.

"Lignes" and "sizes" are the measurements usually used to determine the size of a movement. In Fig. 44 are shown the various diameters of a movement. Of the two main dimensions that of the largest diameter is usually taken, and the most common measurement is the ligne. As 1 ligne equals approximately 3/32 in. a 10 1/2-ligne watch measures 15/16 in. which is short of an inch. The American industry favours the "size" as a unit of measurement. Size O equals 1 5/30 in. Each size above size O increases by 1/30 in., and a size below O decreases by 1/30 in. 10 1/2-ligne movements are to be found in both gentlemen's and ladies' wrist-watches. Until a few years ago this size was almost universal in ladies' watches, and the cheaper watches still favour this size.

FIG. 44.—THE VARIOUS DIAMETERS OF A MOVEMENT.

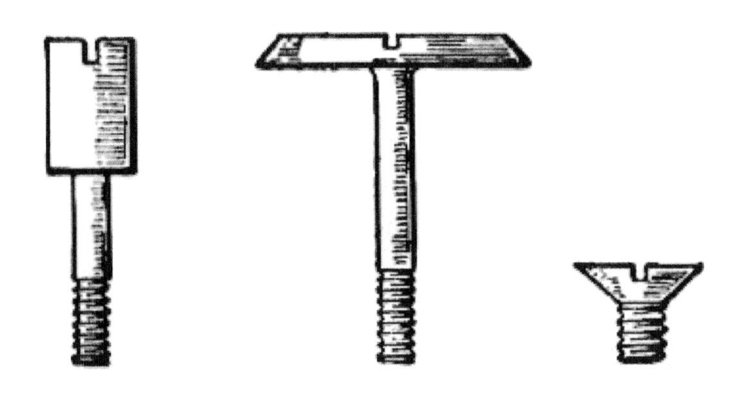

FIG. 45.—THREE DIFFERENT TYPES OF SCREW USED IN THE CONSTRUCTION OF A WATCH.

Number of Screws.—Some watches have as many as 150 separate pieces, and of this large number there are at least 35 screws. Fig. 45 shows three different types of screw: the cheese-headed plate screw, the flat-headed case screw, and the small jewel screw with countersunk head. The main frame of the movement consists of two plates: the bottom or dial plate and the top or back plate, which is visible when the case is opened. The modern back plate has changed from a circular plate into a number of sections usually called bars or bridges, thereby rendering the works easily accessible.

Almost half the movement is used for the large bar that supports the mainspring barrel, a thin cylindrical metal box with teeth around the outside edge. The barrel is fitted with a cover, and the axle upon which it rotates is called the arbor. The arbor has a short hook which engages the inner eye of the mainspring.

The Great Wheel.—The barrel, or main driving wheel, is often referred to as the great wheel, and the other wheels are arranged in the following order. In the centre of the movement and driven by the barrel is the centre wheel; next, the 3rd wheel; then the 4th wheel (the seconds wheel); and finally the 5th wheel (the escape wheel). The remaining section is known as the escapement. When referring to the escapement, this is generally assumed to include the escape wheel, the small anchor-shaped piece called the lever, which arrests and releases the escape wheel tooth by tooth, and the balance and its kindred pieces.

The balance wheel is mounted on a slender axle—the balance staff. Fixed upon the staff above the balance wheel is the hair-spring and below the balance wheel is the roller. The roller is fitted with a small impulse pin, but when a jewelled pin is used it is commonly called the ruby pin. The function of the roller is to unlock the pallets. Fig. 46 shows the balance, balance staff and roller, and the position of the fork of the pallets with regard to the ruby pin.

Jewels.—In jewelled watches, the most popular number of jewels is 15. These jewels are not mere ornaments, but are used to minimise wear. The 15 jewels are always arranged in this order. Two each for the 3rd, 4th, and 5th wheels and pallets, 4 for the balance, 2 pallet stones and the ruby pin. Fig. 49 shows sections of plate and balance jewels. No. 1 is a section of the jewels used for 3rd, 4th, and escape wheels; No. 2 is an endstone; and No. 3 shows the arrangement of the balance jewels, one at each end of the balance staff. In high-class watches, jewel hole and endstone are fixed in separate settings and kept in position by 2 jewel screws as shown. It will be observed that the balance jewel hole differs slightly from the ordinary jewel hole. For example, the oil sink is inside on the balance hole and outside on the plate hole.

The pivots, the short projections of the pinions and staffs which actually rotate in the bearings also differ in shape. In Fig. 48 are depicted at *A* an ordinary pivot with a square shoulder and at *B* a balance pivot with a conical shoulder. Type *A* pivots are used with No. 1 type jewels. In high-class watches the lever and escape wheel pivots are often made conical and provided with balance type jewels and endstones.

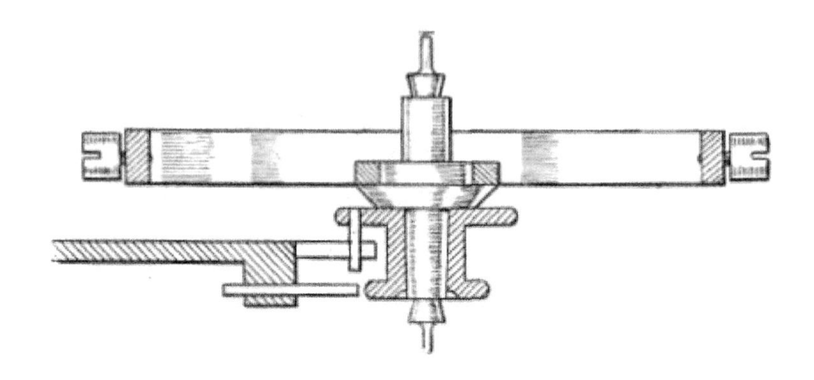

FIG. 46.—THE BALANCE, BALANCE STAFF AND ROLLER, AND THE POSITION OF THE FORK OF THE LEVER WITH REGARD TO THE RUBY PIN.

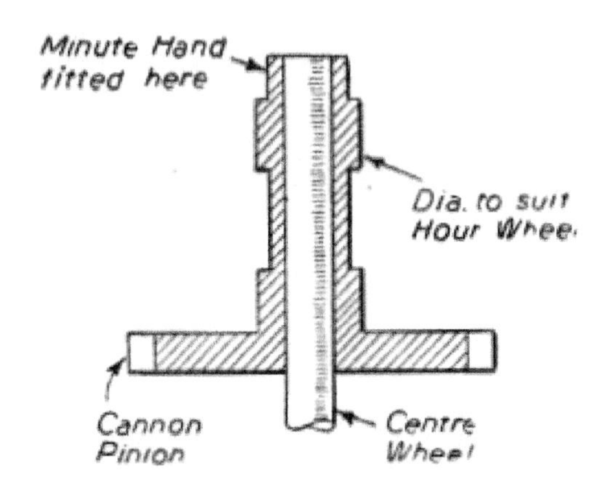

FIG. 47.—A CANNON PINION.

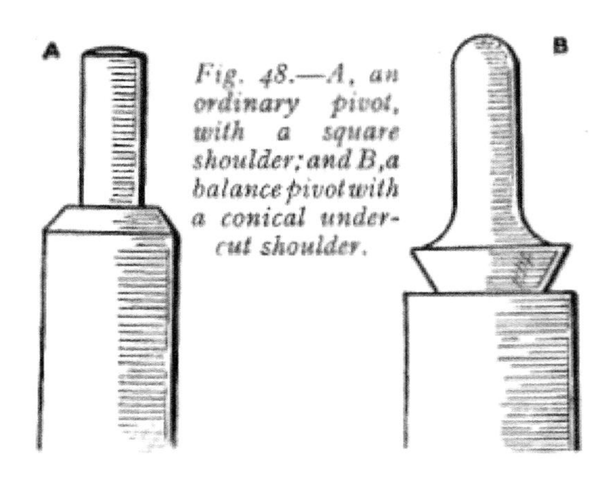

FIG. 48.—A, AN ORDINARY PIVOT, WITH A SQUARE SHOULDER; AND B, A BALANCE PIVOT WITH A CONICAL UNDER-CUT SHOULDER.

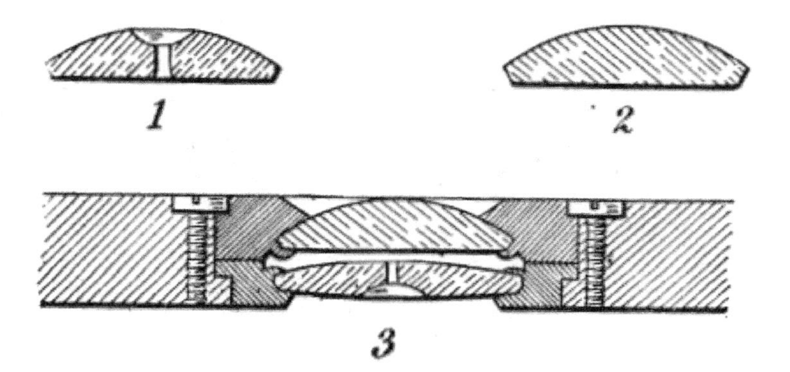

FIG. 49.—SECTIONS OF PLATE AND BALANCE JEWELS.

The Bottom Plate.—Fig. 50 shows a bottom plate. This carries the small winding and hand-setting wheels and the levers that operate them. The winding shaft passes through two small wheels seen at the top, the top or crown wheel engages the smaller of the 2 flat steel wheels seen in Fig. 43 at right-angles. When winding, the mainspring is prevented from "running back" by the action of the pawl or click which arrests the larger of the winding wheels.

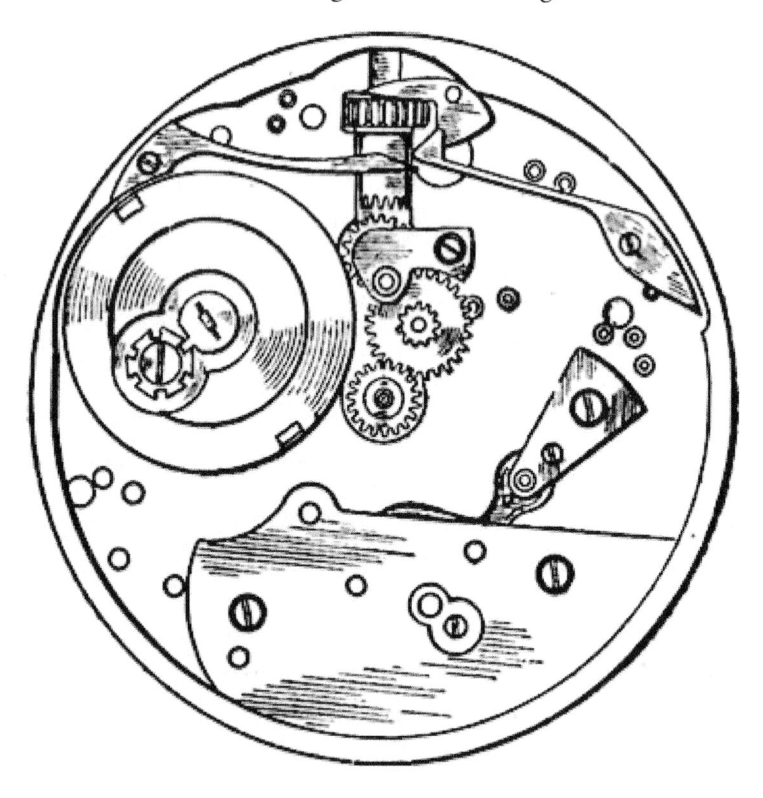

FIG. 50.—A BOTTOM PLATE SHOWING THE PARTS

BEHIND THE DIAL. THIS MOVEMENT IS A PATEK PHILLIPPE.

The winding shaft is prevented from being pulled right out by the pull-up piece, which serves the dual purpose of retaining the winder and forcing down, by means of the return lever, the lower wheel on the winding shaft, causing it to engage the intermediate hand setting wheel. The intermediate wheel gears with the minute wheel—a flat brass wheel having a short pinion and rotating on a stud fixed in the plate—and the teeth of the minute wheel gear with those of the cannon pinion. The cannon pinion is really a small tube with teeth around the bottom and it fits friction tight on the extended pivot of the centre wheel. It is upon this tubular pinion that the minute hand is fixed.

When the winder is pulled out, it depresses the lever and the castle wheel, and the motion wheels (the hand wheels), are engaged and can be turned around to the desired position. Fig. 47 shows the cannon pinion.

THE COMPENSATING BALANCE & HAIRSPRING

To illustrate the general principles of timing, it may be of interest first to make a comparison between the balance in a watch and the pendulum in a clock, as both of them evidently perform the function of measuring or beating time. The pendulum, as we all know, requires no special spring to bring it to its centre line, the perpendicular, as the force of gravity furnishes the necessary power for doing this work in a very ideal way. When a pendulum is put in motion, it makes a vibration in a certain interval of time, and in proportion to its length, regardless of its weight, because the force of gravity acts on it in proportion to its mass. The length of a pendulum is reckoned from its centre of suspension to its centre of oscillation, which latter point is located a short distance below the middle of the bob. If a weight is added *above this point*, the clock will *gain*, because it raises the centre of oscillation and has the same effect on the time-keeping as raising the whole bob, which is equivalent to a shortening of the pendulum; but if a weight is added *below this point, it has the opposite effect*, as it really lengthens the pendulum. Reasoning from these facts we come to the conclusion that we can make a certain change in the rate of a clock in three different ways. For example, we may make it gain: (1) by raising the bob; (2) by adding weight above the centre of oscillation; and (3) by reducing the weight below that point. An interesting fact in relation to the pendulum, which may not be generally known among watchmakers, is that its rate of vibration varies slightly with change of latitude, and also of altitude (that is, its height above the sea-level), making a clock lose at the Equator and at high altitudes, and gain as we go nearer the sea-level and the Poles. This is due, partly to the distance from the centre of the earth, which is greater at the Equator than at the Poles, and partly to the centrifugal force resulting from the rotation of the earth on its axis. Both these factors tend to make an object weigh less (on a spring balance) at the Equator than at the Poles, and also cause a change in the rate of a clock as stated above. In view of these facts we might, as a fourth way of making a clock gain—although not a very practical one—move to a locality nearer the Pole. A balance is different from a pendulum in three fundamental points: first, it is poised; consequently the force of gravity has no effect on it, except as its influences the friction on its pivots; second, the vibrations are controlled by a spring instead of the force of gravity; third, a weight (mass) added to a balance will always make it vibrate slower, provided it is not thereby put out of poise, and the retarding effect will be greater the farther the weight is placed away from its centre. One difficulty encountered in the first attempt to make accurate timepieces was the variation in the dimensions of metals caused by difference in temperature. All metals with the exception of a recently discovered alloy of steel and nickel (64 parts of steel and 36 of nickel) have the property of expanding with increase of temperature—the different metals showing a somewhat different rate of change. As the length of the pendulum is the all-important factor in the timing of clocks, so also is the diameter of the balance and the length and resiliency of the hairspring in a watch. It is absolutely necessary to devise some means of compensating for changes in temperature before a reliable timepiece of either form can be made. So far as this problem applies to clocks, the mercury pendulum proves to be a very satisfactory solution, at least so far as accuracy is concerned. The bob of this pendulum is composed of one or more tubes of glass or iron, and these tubes are filled with mercury to a certain height. When of proper dimensions, the expansion and contraction of this column of mercury raises or lowers its mass to exactly compensate for the change in the length of the pendulum rod due to variations in the temperature. This method, although very satisfactory for clocks, cannot, of course, be applied to watches, for obvious reasons, but for this purpose we make use of the property of the metals alluded to above, namely, *the difference* in the ratio of *expansion in different metals.*

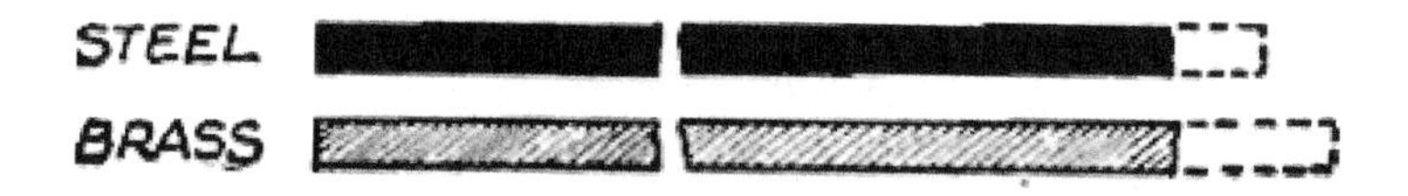

FIG. 51.—TWO BARS OF EQUAL LENGTH WHEN AT NORMAL TEMPERATURE. THE DOTTED LINES INDICATE THE RELATIVE EXPANSIONS WHEN EACH IS HEATED TO A SIMILAR DEGREE.

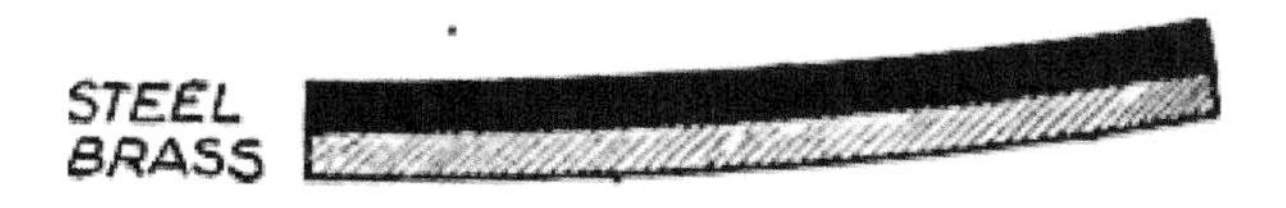

FIG. 52.—THE EFFECT OF HEAT ON THE BI-METALLIC BAR. THE GREATER EXPANSION OF THE BRASS CAUSES THE BAR TO CURVE UPWARD.

FIG. 53.—THE TWO METALS FUSED TOGETHER.

FIG. 54.—IF COLD WERE APPLIED INSTEAD OF HEAT, THE BAR WOULD CURVE IN THE OPPOSITE DIRECTION.

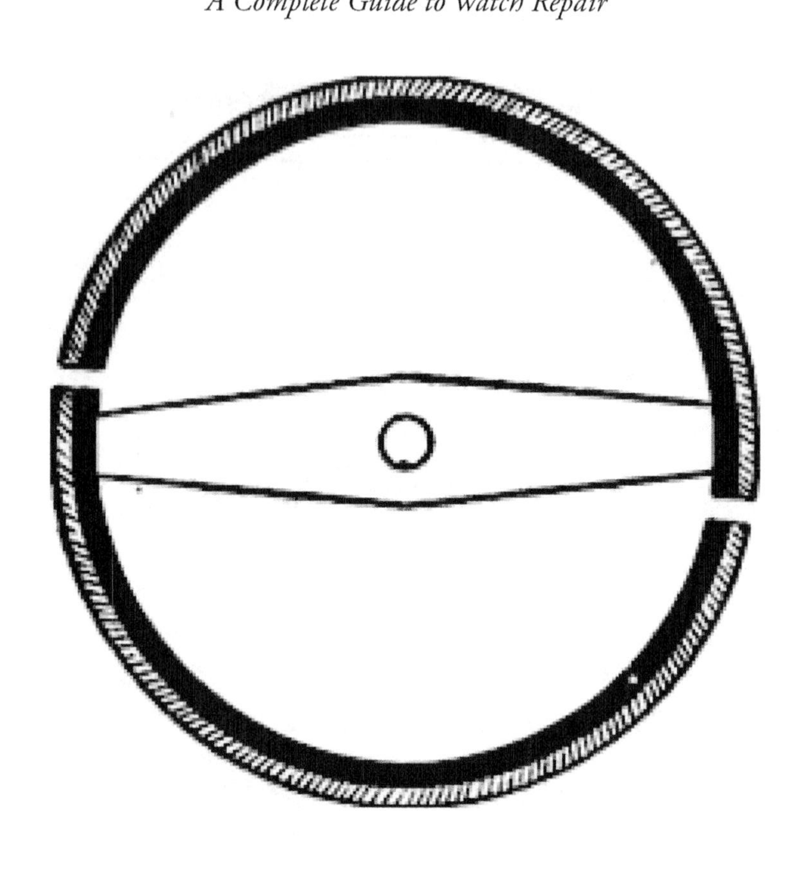

FIG. 55.—WATCH BALANCE IN WHICH THE ARMS AND INNER PORTION OF THE RIM ARE OF STEEL. THE OUTER PORTION IS BRASS FUSED TO THE STEEL. THE RIM IS SEVERED AT TWO POINTS NEAR THE ARM, PERMITTING THE RIM TO MOVE UNDER CHANGE OF TEMPERATURE.

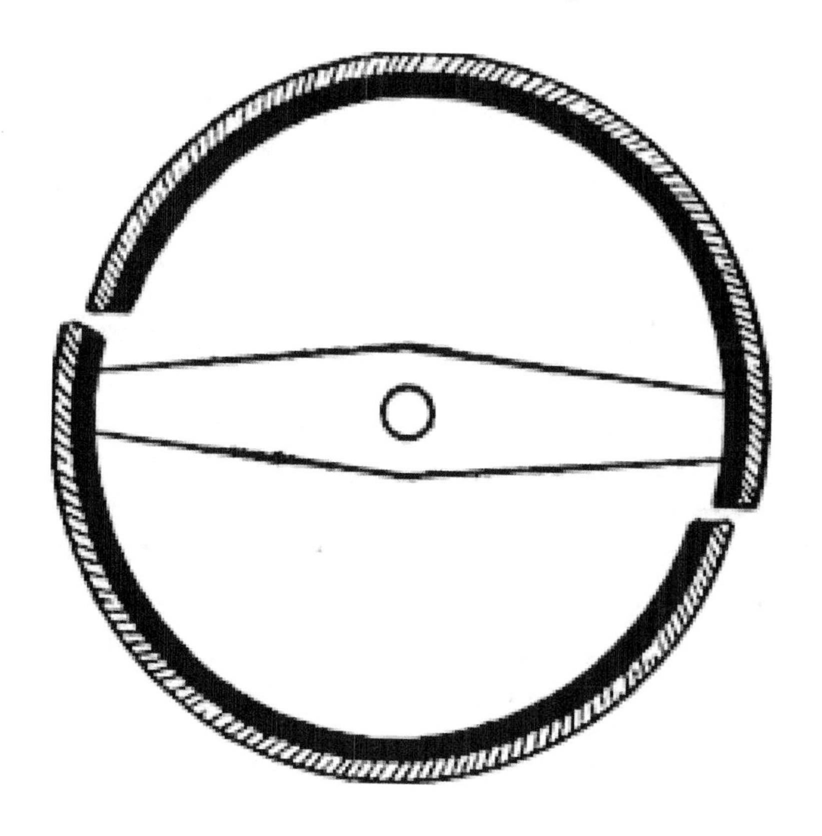

FIG. 56.—THE BALANCE UNDER THE INFLUENCE OF HEAT. IT WILL BE NOTICED THAT THE FREE ENDS OF THE BALANCE RIM HAVE CURVED INWARDS, THUS REDUCING THE DIAMETER OF THE BALANCE (RADIUS OF GYRATION).

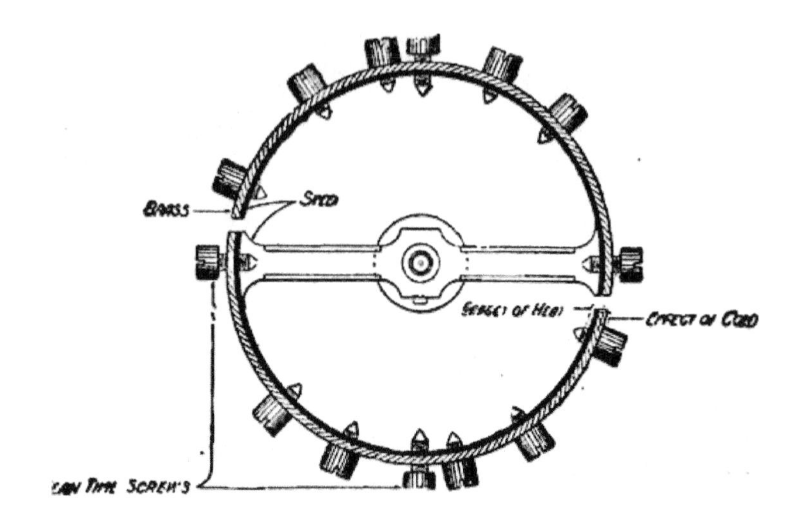

FIG. 57.—DIAGRAM SHOWING THE BALANCE, LOADED WITH SCREWS FOR TIMING AND ADJUSTMENT AND COMPENSATION.

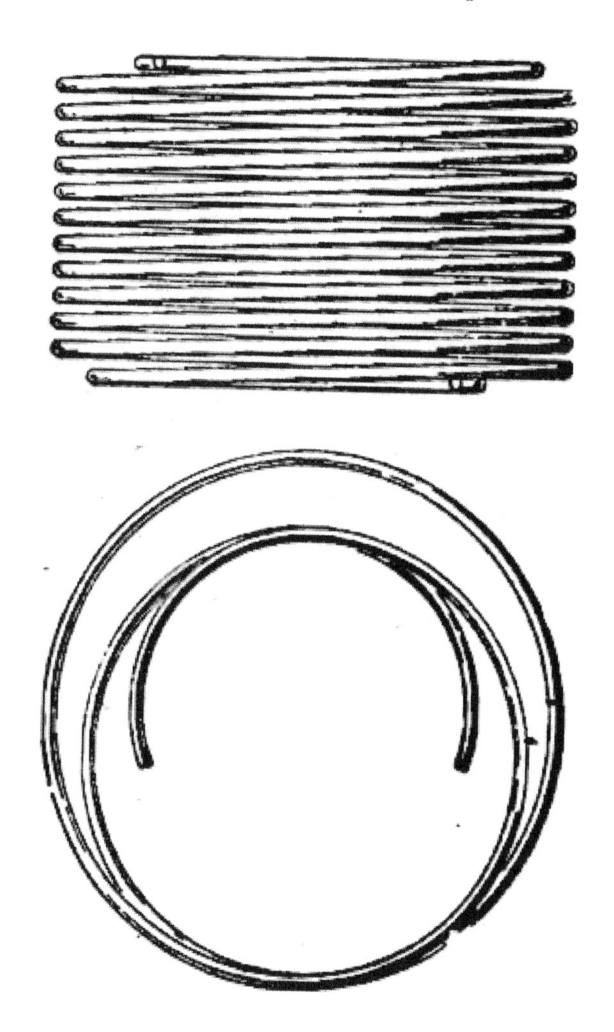

FIG. 58.—HELICAL BALANCE SPRING FITTED USUALLY TO MARINE CHRONOMETERS.

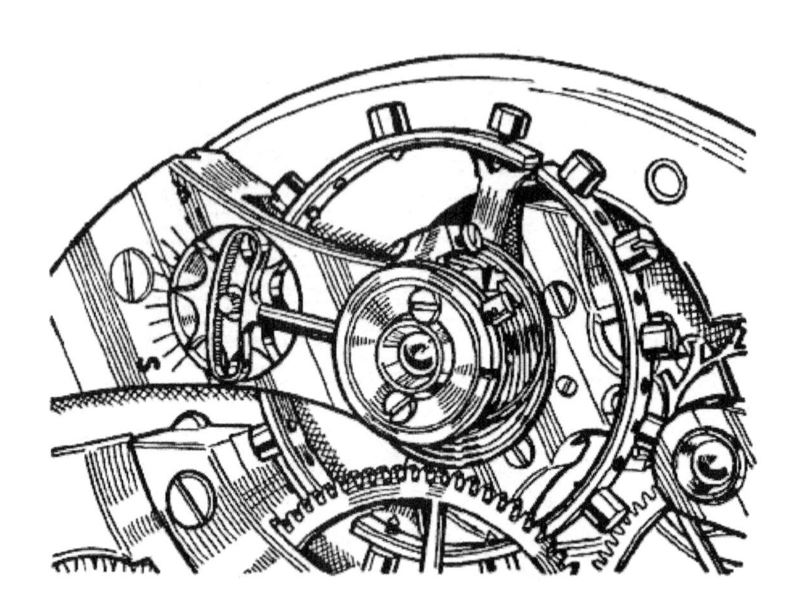

FIG. 59.—THE WALTHAM MICROMETRIC REGULATOR. OTHER MAKERS USE AN EXTERNAL RACK AND PINION, OR A SPRING-LOADED CAM.

FIG. 60.—AN UP-TO-DATE BI-METALLIC MARINE CHRONOMETER BALANCE, MADE FROM BRASS AND STEEL.

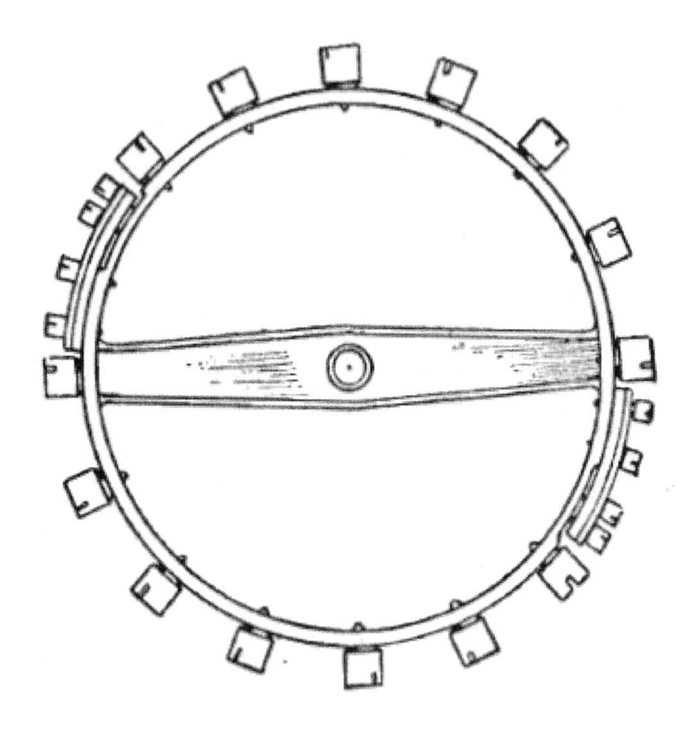

FIG. 61.—THE "AFFIX" DITISHEIM MONO-METALLIC UNCUT BALANCE. COMPENSATION IS SECURED BY THE TWO SMALL AUXILIARY PIECES, AND THE USE OF AN ELINVAR HAIRSPRING.

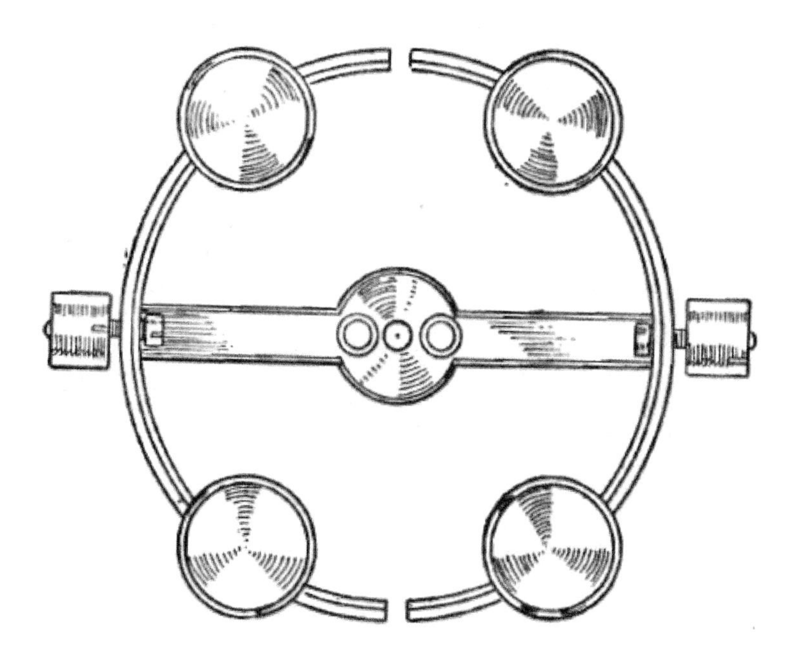

FIG. 62.—THE GUILLAUME BALANCE FOR CHRONOMETERS. IT IS, OF COURSE, BI-METALLIC AND CUT. MADE FROM BRASS AND NICKEL STEEL, IT ALMOST GETS RID OF MIDDLE TEMPERATURE ERROR.

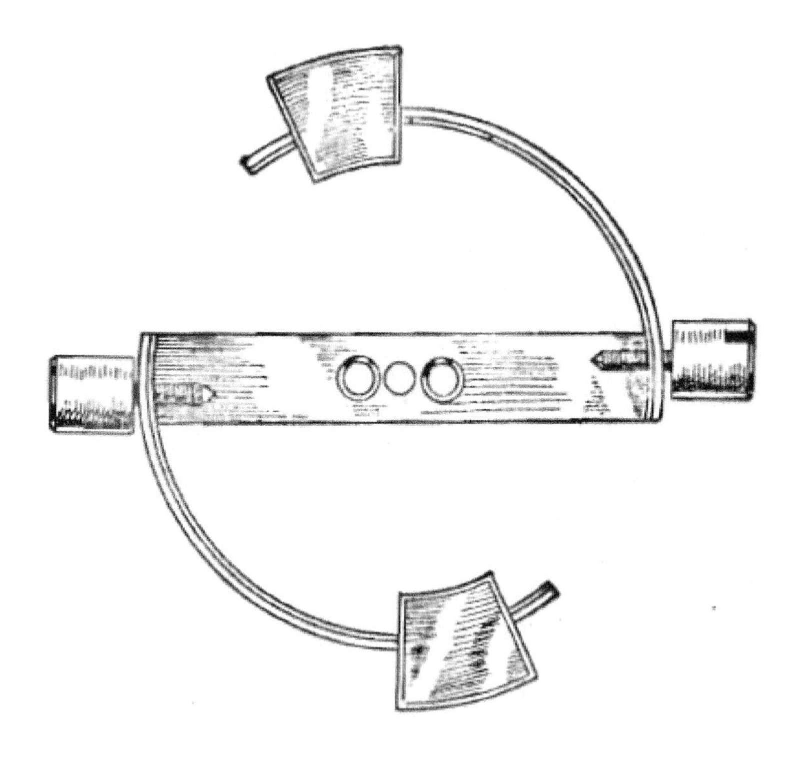

FIG. 63.—MARINE CHRONOMETER BI-METALLIC BALANCE AS INTRODUCED BY EARNSHAW.

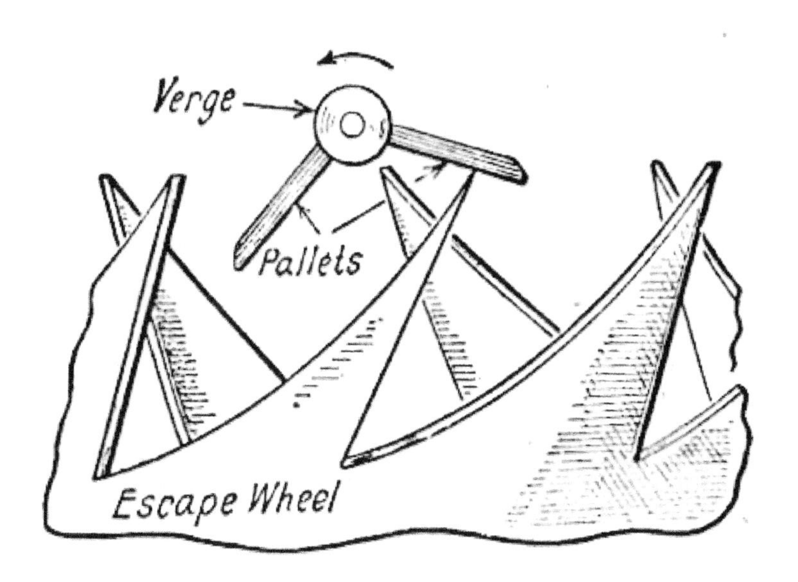

FIG. 64.—THE VERGE ESCAPEMENT—NOW OBSOLETE.

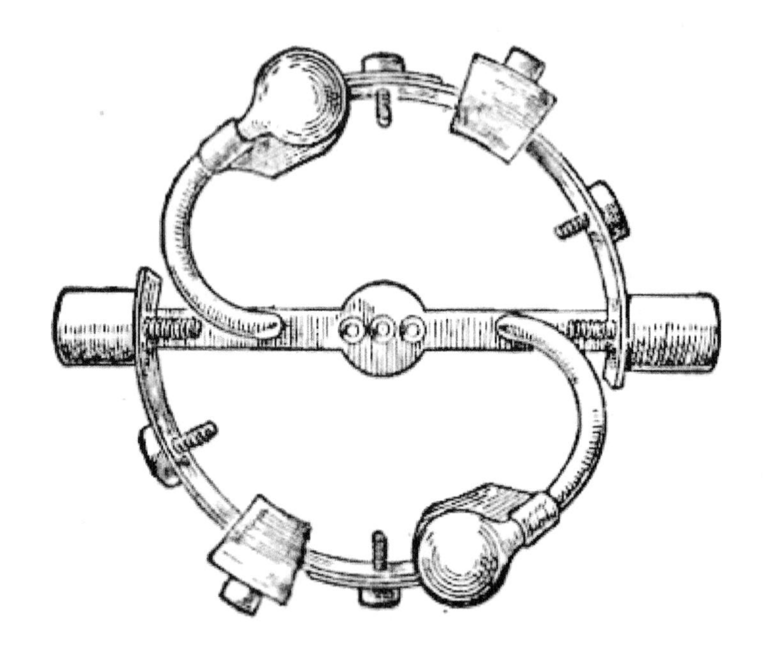

FIG. 65.—THE LOSEBY COMPENSATING BALANCE; ON EACH ARM OF THE RIM IS A VESSEL CONTAINING MERCURY.

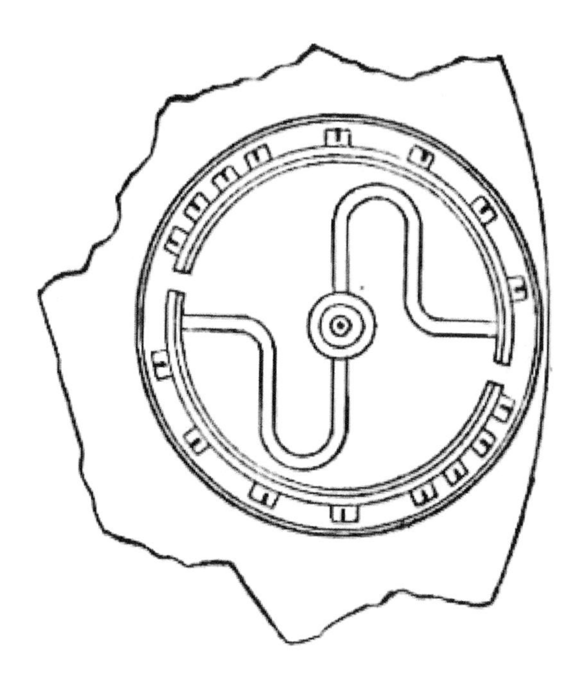

FIG. 66.—THE WYLER COMPENSATING BALANCE AS FITTED TO THE WYLER SELF-WINDING WRIST-WATCH. IT IS SURROUNDED BY A PROTECTING RIM, AND IS SHOCKPROOF.

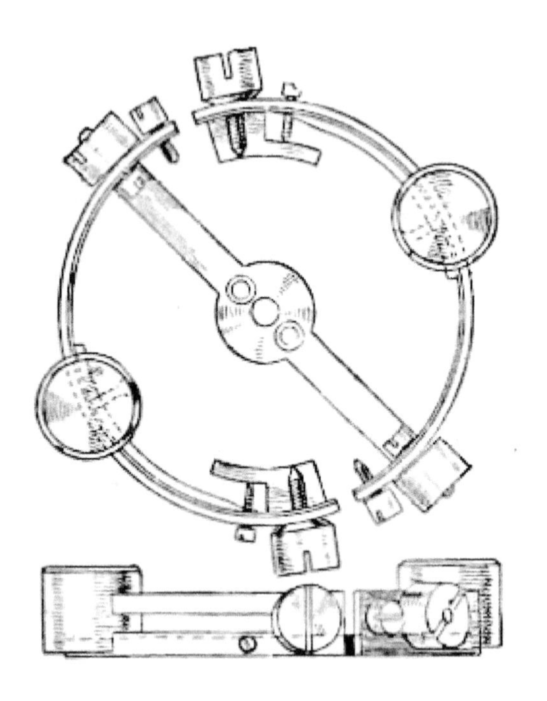

FIG. 67.—THE AUXILIARY COMPENSATION BALANCE INVENTED BY KULLBERG. THE RIM IS OF THE USUAL BI-METALLIC TYPE. THE STEEL FOR 80° ON EACH SIDE OF THE BALANCE IS MADE THICKER. ITS OBJECT IS TO OVERCOME MIDDLE TEMPERATURE ERROR. USUALLY FITTED TO MARINE CHRONOMETERS.

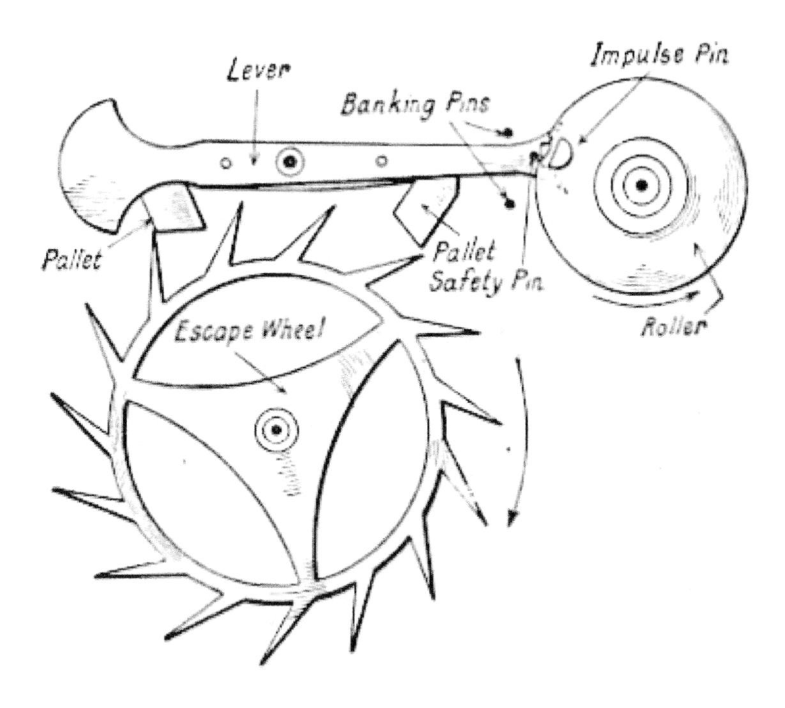

FIG. 68.—THE ENGLISH OR SIDE LEVER. THIS FORM OF ESCAPEMENT IS RAPIDLY BEING SUPERSEDED BY THE "STRAIGHT LINE" ESCAPEMENT, IN WHICH THE CENTRES OF THE BALANCE, LEVER AND ESCAPE WHEEL ARE IN LINE. THE SPUR TOOTH ESCAPE WHEEL HAS ALMOST ENTIRELY BEEN REPLACED BY THE CLUB TOOTH ESCAPE WHEEL MADE OF STEEL, NOT BRASS AS IS THE SPUR WHEEL.

A number of holes are drilled radially through the bi-metallic rim, and these holes are tapped to receive the balance screws. Usually about twice as many holes are made in the rim as the number of screws used in the balance; this is done to give opportunity for moving the screws in the final adjusting to temperatures. The object in using screws in the balance rim is two-fold; first, to provide the necessary weight (mass) in the rim, and second, to have this weight movable for temperature adjustments, as stated above.

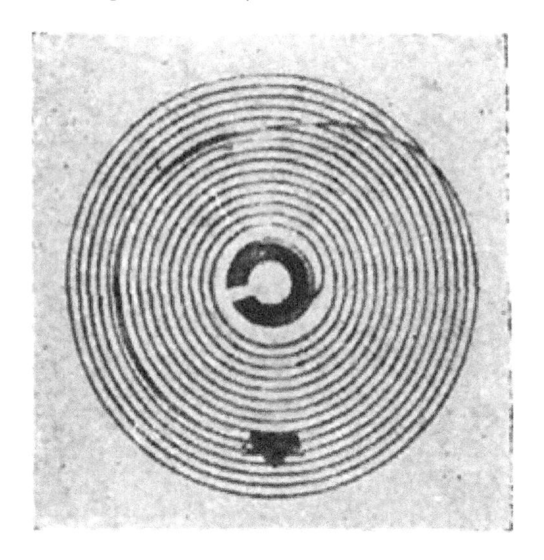

FIG. 69.—THE OVERCOIL HAIRSPRING, THE OUTER CURVE OF WHICH IS MADE TO CONFORM WITH PHILLIP'S THEORY. IT GIVES A CONCENTRIC ACTION TO THE HAIRSPRING AND THUS REDUCES POSITION ERRORS. SUCH AN OVERCOIL IS OFTEN

ERRONEOUSLY CALLED "BREGUET," WHICH IS A SPECIALLY FORMED CURVE. IT IS IMPOSSIBLE TO ELIMINATE POSITION ERRORS WITH A FLAT HAIR-SPRING.

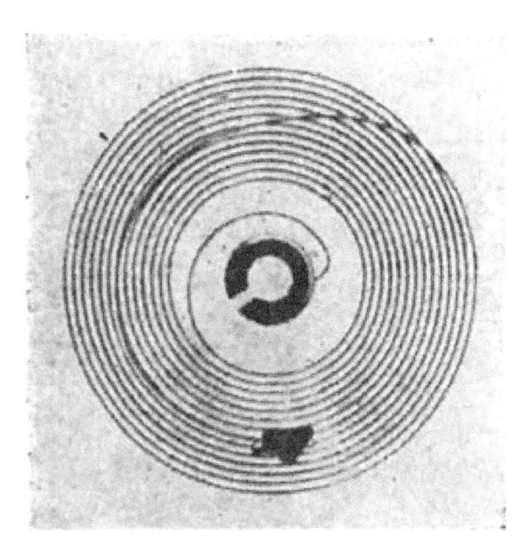

FIG. 70.—THE WALTHAM HAIRSPRING SHOWING INNER AND OUTER TERMINAL CURVES. THE INNER CURVE SHOWN IN FIG. 69 TENDS TO THROW THE SPRING OUT OF POISE DURING ITS VIBRATIONS. WITH THE INNER CURVE SHOWN ABOVE, THIS TENDENCY IS ELIMINATED. WALTHAM HAIRSPRINGS ARE HARDENED AND TEMPERED IN FORM. MOST OTHER SPRINGS ARE FORMED AFTER THE SPRINGS HAVE BEEN TEMPERED.

We will now understand from what has been said that when a compensating balance is exposed to a higher temperature, every part of it expands, or grows larger, but as a result of the combination of the two metals in the rim, and the ends of the rim being free to move, each half of the rim will curve inward, carrying its weight towards the centre of the balance, and thus compensate for the lengthening of the arms and the weakening of the hairspring. If a balance is exposed to a lower temperature, the action will, of course, be in the opposite direction.

FIG. 71.—PLAN AND ELEVATION OF THE TOURBILLON ESCAPEMENT. THE BALANCE IS MOUNTED ON A CARRIAGE WHICH ITSELF REVOLVES ONCE EVERY MINUTE, THUS ELIMINATING POSITION ERRORS. IN THIS DIAGRAM A IS THE FOURTH WHEEL, B THE REVOLVING CAGE OR CARRIAGE, D CARRIAGE PINION, E THE ESCAPE PINION, F PIVOT FOR SECONDS HAND, H ESCAPE COCK, G UPPER PIVOT, C THE THIRD WHEEL GEARING WITH D. THE KARRUSEL, INVENTED BY BONNIKSEN, IS SOMEWHAT SIMILAR EXCEPT THAT IT ROTATES ONCE IN 52 1/2 MINUTES.

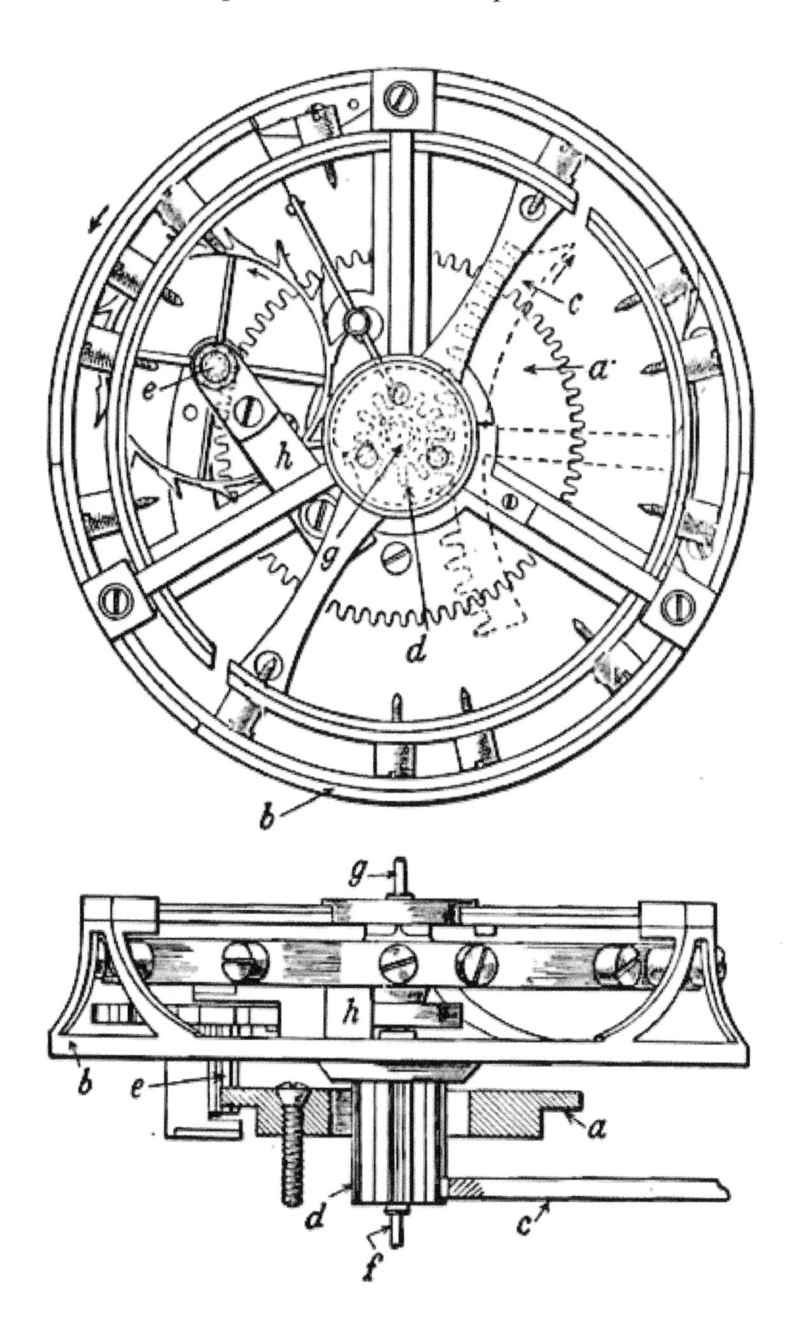
c
a
e
h
g
d
b
g
h
b
e
a
d
c
f

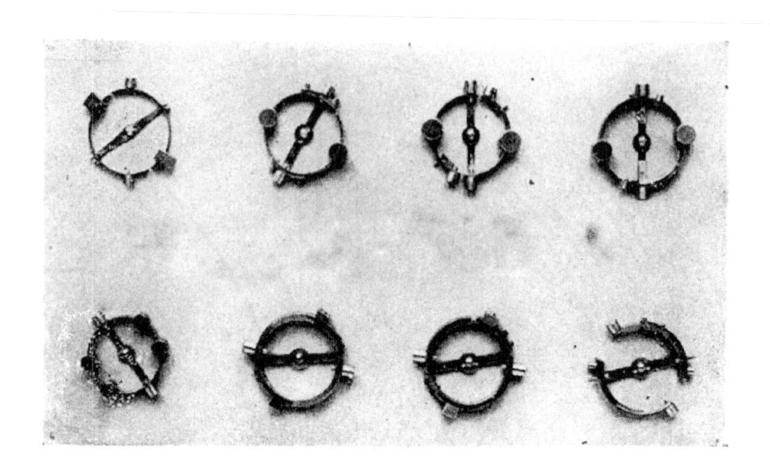

FIG. 73A.—EIGHT EXAMPLES OF COMPENSATING BALANCES FOR MARINE CHRONOMETERS.

When a watch is to be adjusted to temperatures, it is run 24 hours, dial up, in a temperature of 90°F., and its rate compared with a standard. It is then run 24 hours, dial up, in a temperature of 40°F. If its shows a gain in the 40° temperature, as compared with the running in the 90°, it is said to be under - compensated. This is remedied by moving some screws nearer the free ends of the rim. This will, of course, result in a greater compensating effect, because the screws which we move nearer the ends of the rim must travel a greater distance in or out in relation to the centre of the balance when the balance is exposed to changes of temperature. After the screws have been moved, the movement is tried again the same length of time, and so on, until it runs the same in both temperatures. When a screw is moved in one side of the balance, it is, of course, necessary that the corresponding screw in the other side should be moved the same. A modern compensation balance, combined with a correctly proportioned steel Breguet hairspring, which has been hardened and tempered in form, constitute a time-measuring device of marvellous accuracy. And the bi-metallic rim, hardened as the Waltham balances are, so as to be perfectly safe against distortion from ordinary handling, is certainly a boon to the watchmaker.

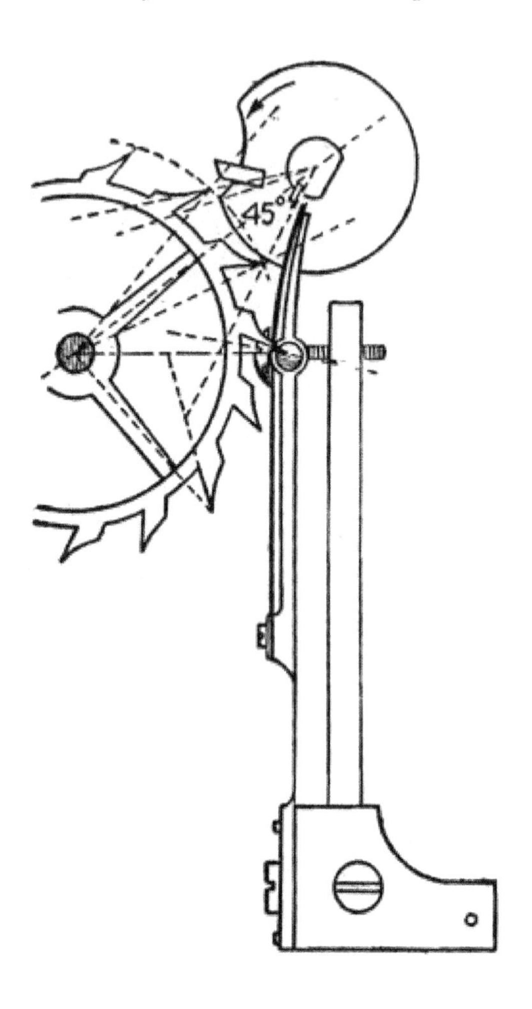

FIG. 72.—ANOTHER TYPE OF MARINE CHRONOMETER ESCAPEMENT, AS DESIGNED BY EARNSHAW. THE DISADVANTAGE IS THAT IT TENDS TO "SET" OR STOP WHEN WORN IN THE POCKET.

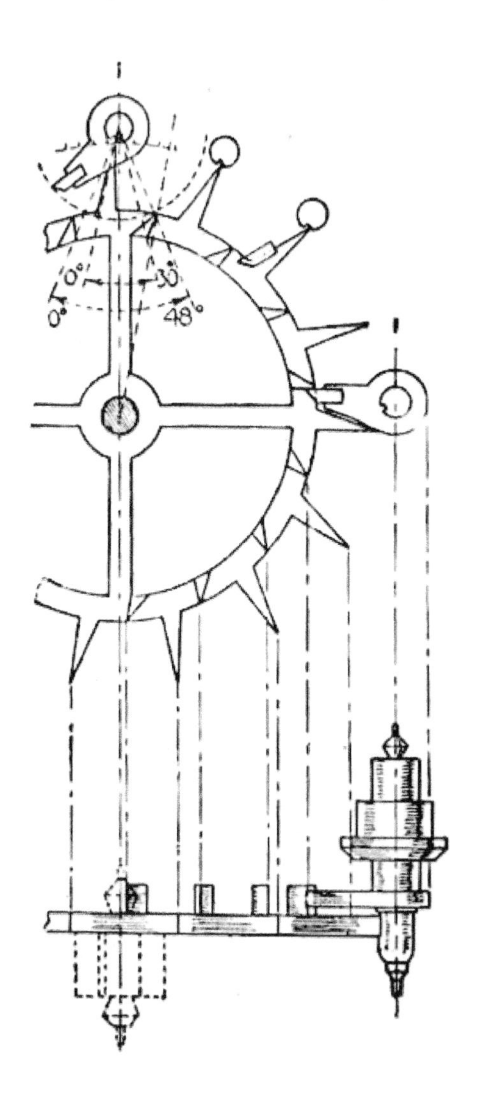

FIG. 73.—THE DUPLEX ESCAPEMENT. THIS TYPE WAS EXTREMELY POPULAR AT ONE TIME, AND WAS FINALLY FITTED TO THE CHEAP WATER-BURY WATCHES. IT IS NOW OBSOLETE.

The mean time screws used in the Waltham balances furnish an excellent means for accurate timing, as two, on opposite sides, can be turned an equal amount in (making the watch run faster) or out (slower) without changing the poise of the balance.

The following is the approximate effect of *one-half turn* of two mean time screws:

18-size and 16-size	2½ seconds per hour	
Colonial series and 12-size ..	2½ ,, ,,	
0 size and 3-0-size	3 ,, ,,	
Jewel series	3½ ,, ,,	
10 Ligne	3½ ,, ,,	

An illustration of a hairspring supplied with a theoretically correct outer terminal, commonly known to watchmakers by the name of Overcoil, is shown in Fig. 69.

As is well known to watchmakers, hairsprings are supplied with overcoils to secure a concentric action of the hairspring while the balance is in motion. A concentric action of the hairspring is necessary, in order to reduce the position error.

This result is partially obtained when a hairspring is supplied with a theoretically correct outer terminal or overcoil, whereby the centre of gravity of the outer half of the hairspring is made to coincide with the centre of the balance at every stage of its vibration.

In a watch fitted with a hairspring with an outer terminal curve only, there still remains a position error, part of which, at least, is due to the fastening of the inner end of the spring, which, in its ordinary form, tends to throw the spring out of poise during its vibrations.

This the Waltham Watch Company has succeeded in overcoming by making hairsprings with theoretically correct inner terminal curves. This inner curve maintains the body of the spring in perfect poise with the balance, during both its opening and closing vibrations. I show an illustration (Fig. 70) of a hairspring of this kind with inner and outer terminals.

The design of the hairspring having been perfected, there came the problem of properly producing these springs.

Flat hairsprings, that is to say, those without an overcoil, are only fitted to cheaper watches, although high-grade thin dress watches are sometimes fitted with them in order to reduce the thickness of the movements. It is impossible to eliminate position errors when a flat hairspring is fitted.

THE LEVER ESCAPEMENT

The proper action of the human heart is no more essential to ensure a sound and healthy body, than is the proper adjustment and action of the escapement to the reliable performance of the watch. If the watch escapement is properly made and adjusted it will not only run—but will run with marvellous accuracy. So the timekeeping qualities of the watch are in large measure dependent on the condition of the escapement. It is therefore of great importance that every watchmaker should acquire an intimate knowledge of all the actions that are involved in the kinds of escapements with which he has anything to do.

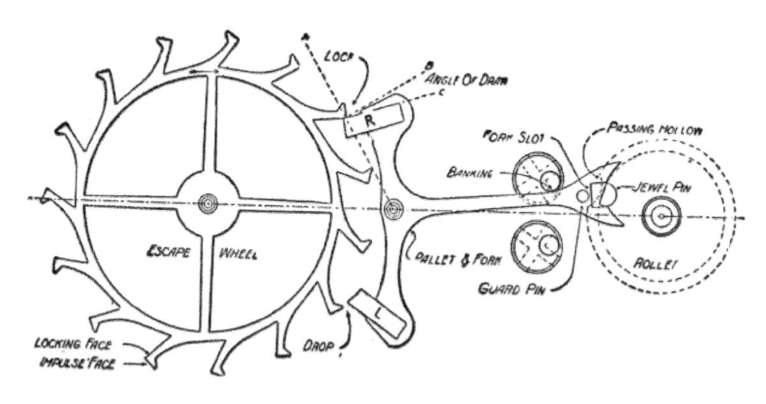

FIG. 74.—ALL GOOD WATCHES HAVE A LEVER ESCAPEMENT OF THE TYPE SHOWN HERE. NOTE THE JEWELLED PALLETS.

The only kind of escapement used in modern watches is the detached lever escapement, sometimes designated as the anchor escapement. This escapement requires no special introduction to watchmakers, for by extensive use, and by the test of time, it has been proved to be the most practical, as well as the most reliable form of escapement for pocket timepieces.

I include in this chapter some drawings of the lever escapement that they may be convenient for reference, and an aid to the clear understanding of the text.

The function of the escapement is to impart to the balance, regularly, and with as small loss as possible, the power which has been transmitted through the train from the mainspring to the escape pinion. In the lever escapement this is accomplished by means of two distinct actions: first, the action of the escape wheel and pallet; second, the action of the fork and roller pin. Fig. 74, is a plan view of the Waltham Lever Escapement, as used in the 16-size, and the 18-size models; movements drawn to scale 5-to-1, and giving the names of the principal parts and features of the same. The escape wheel is mounted friction tight on the slightly tapered staff of the escape pinion. It has 15 teeth, called "Club Teeth" on account of their peculiar shape, resulting from the addition of impulse faces to the ends of the teeth, and to distinguish them from "ratchet teeth", the name given to a style of pointed teeth used on escape wheels in an earlier form of lever escapement. In descriptions of this escapement the term "exposed pallets" is used. This means that the pallet stones are visible, with the active ends standing out free from the body of the pallet, as distinguished from an earlier form of pallet with "covered stones" set in slots running in the plane of the pallet.

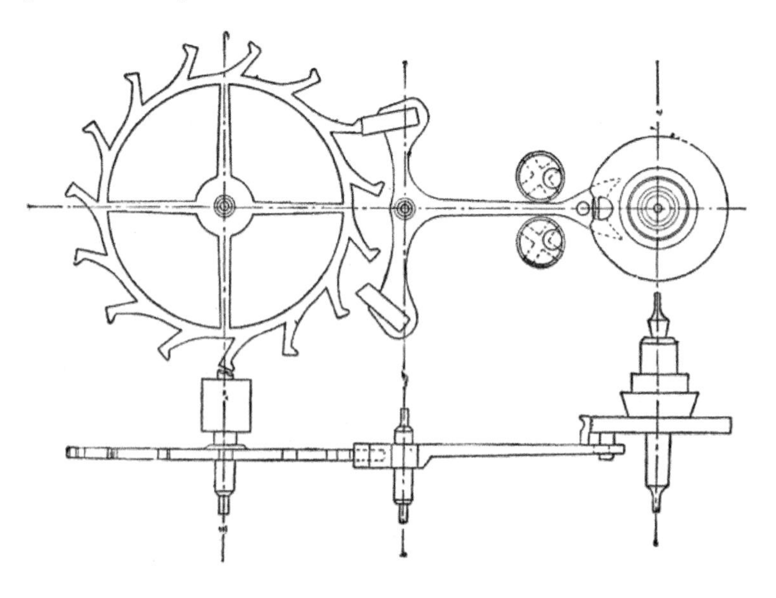

FIG. 75.—LEVER ESCAPEMENT OF THE SINGLE ROLLER TYPE.

The Pallet Action.—The action of the escape wheel and pallet includes the following features: impulse, drop, lock, draft and slide, and in giving a general description of these actions we will consider briefly what constitutes each one of these features. The pallet is of the kind called "circular pallet", which means that the distance from the pallet arbor to the middle of the impulse face is the same for both pallet stones. Another kind of pallet is made with "equidistant lock", that is, the distance from the pallet arbor to the point where the lock takes place, is the same for both pallet stones. The pallet is mounted on its arbor, which is located close to the periphery of the escape wheel. A theoretically correct distance in relation to the diameter of the escape wheel will not allow an excess of clearance between the pallet and the escape wheel teeth when opposite the pallet arbor, and for that reason the amount of stock in the pallet is made very small at that point. The pallet is slotted for the 2 pallet stones in such a way as to make the inside corners of the pallet stones reach over 3 teeth of the escape wheel, and to make the outside corners of the stones reach over 2 teeth and 3 spaces of the wheel, with a small amount of clearance in each instance which is called the "drop".

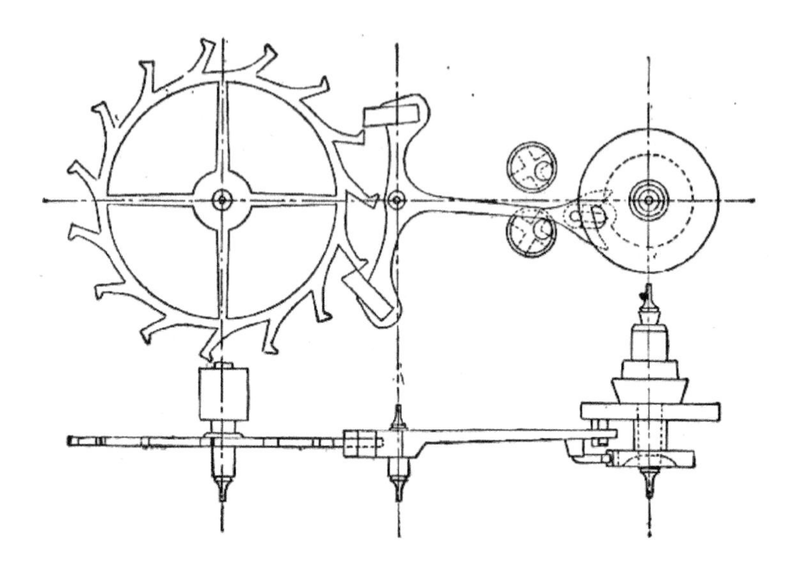

FIG. 76.—LEVER ESCAPEMENT OF THE DOUBLE ROLLER TYPE.

One other important point in relation to the slotting is to direct the slots in the pallet in such a way as to make the locking faces of each pallet stone present to the locking corners of the escape wheel teeth a certain angle of "draw" when the stones are in the position of "lock". Suppose that the escape wheel is being forced in the direction indicated by the arrow, but is prevented from turning in that direction because the locking face of the *R* pallet stone is directly in the way of a tooth. The particular tooth which is resting on the pallet stone is exerting a certain pressure directly towards the pallet arbor. If the locking face of the pallet stone were along the line *B*, which is at a right-angle to that line of pressure, there would be no tendency for the pallet to turn in either direction, but being along the line *C*, which forms an inclined plane in relation to the direction of the pressure, the pressure applied by the escape wheel tooth will tend to pull the pallet stone towards the escape wheel. This action is called the "draft" or "draw". The turning of the pallet is, however, limited by the banking pin, and the object of the draw is to keep the fork against the banking pin all the time that it is not in engagement with the jewel pin. This action of "draw" is similar on the *L* stone; the only differences are, first, that the pressure of the escape wheel tooth is exerted in the direction *away* from the pallet arbor, instead of towards it, and second, that the turning of the pallet, which in this instance is in the opposite direction, is limited by the other banking pin.

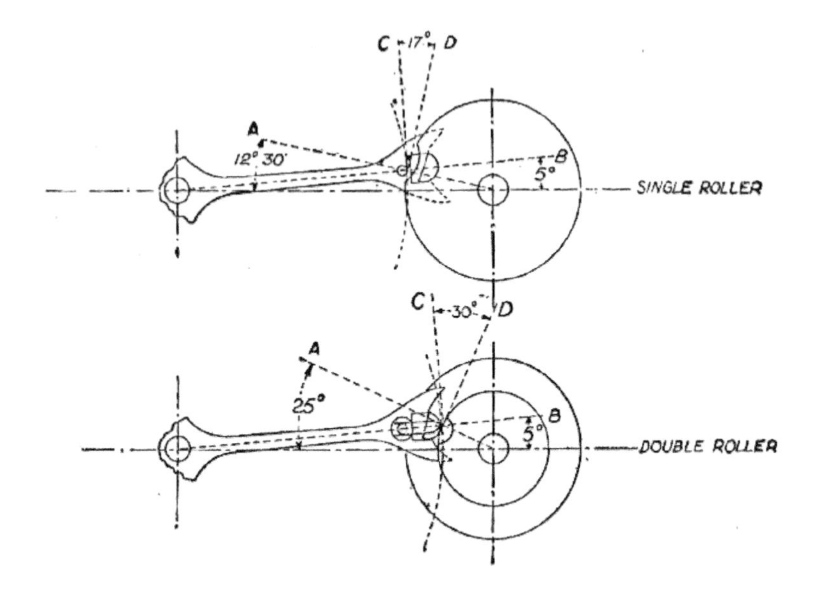

FIGS. 77 & 78.—THE SAFETY ACTION.

A glance at the drawing will make it apparent that the impulse face, which is formed by the surface between the locking and the let-off corners, is at a different angle on the *R* from that on the *L* stone. The impulse angles of the stones in the escapement represented by this drawing are, on the *R* stone, 6° 30', and on the *L* stone 14°. The above refers in each instance to the angle of the impulse face in relation to a right-angle to the locking face, or to the body of the stone. From this condition and from the direction of the pallet stones in relation to the body of the pallet, the factory names "straight" and "crooked" have been given to the *R* and *L* stones, respectively. In books and treatises on the lever escapement the names "receiving" and "discharging" are used, but when, as a matter of fact, both pallet stones perform the functions of receiving and discharging, one as much as the other, these names do not seem to be appropriate. For my part I prefer to use the letters *R* and *L* to distinguish one stone from the other, and these letters may easily be remembered as right and left, this being the order in which the stones appear as we look at the top of the pallet with the fork turned towards us.

The impulse or lift is divided between the escape wheel clubs and the pallet stones; the two together cause the pallet to turn through an angle of 8°30⊠.

The lock amounts to from 1° to 1° 30⊠, making the total angular motion of the pallet about 10°. This is the condition when the pallet is "banked to drop", that is, when the teeth of the escape wheel will just barely pass by the let-off corners of the pallet stones as the fork comes to rest against the bankings. A certain amount of clearance, or freedom, has to be added to this to allow for oil, etc., so that the bankings have to be turned away from the centre line a small amount to allow for what is called "slide", that is, the pallet stone will slide a visible amount into the escape wheel, after the escape wheel tooth drops on to the same. The amount of slide should, however, be very small, because it causes loss of power, by increasing the resistance to unlocking as, in order to un'ock, the escape wheel actually has to be turned backwards against the power of the mainspring, and the amount of this recoil is in proportion to the lock and slide added together. It is therefore important to notice the action of every tooth of the escape wheel on both pallet stones, to ascertain that each pallet stone has some slide on every tooth, and to allow only a small amount in the place where it appears to be closest.

Roller and Jewel Pin.—One problem in connection with lever escapements, with which every watchmaker has had more or less experience, has come to a final solution in the modern double roller escapement. This is the fastening of the jewel pin. The roller, which holds the jewel pin, is made of bronze, with a hole in it the shape of the pin, but a certain amount smaller than the pin. The jewel pin is made of sapphire, and is made slightly tapering, and is forced into the hole in the roller, thus making it permanently secure. The shape of the jewel pin is round, with one side flattened off to measure three-fifths of the diameter of the pin, and the sharp corners removed. This form of jewel pin is superior in general practice to any other form, as it unites strength with the most desirable shape at the points of action. The principle of setting the jewel pin directly in the roller, without cement, is made possible by the double roller escapement, because of the special roller for the safety action. It would not be practical to set jewel pins without cement in steel rollers, as it is in bronze, neither would it be advisable to use bronze rollers for the safety action, because it has been found by experience that tempered steel is better for that purpose. But by separating the two features, it makes a most desirable combination to use a bronze roller for carrying the jewel pin, and a separate steel roller for the safety action.

Matching the Escapement.—The term "matching the escapement" is used to designate the work of bringing the different parts of the escapement into correct relation to each other; in other words, to make the necessary moves in order to obtain the proper lock, draft, drop, slide, fork length, let-off, etc. The best way of learning to do this work is to have a competent instructor who is at hand ready to inspect and to give advice. The difficulties are not so great in *doing* this work, as in correctly determining *what* to do, in order to bring about certain results, and also to know when the escapement is in a proper condition. It is difficult to give in writing a comprehensive idea of how to do this work. I will, however, give a few points which I hope will be useful to the beginner.

The first thing to receive attention is the condition of the pivots on the escape pinion, pallet arbor and balance staff, to see that they are straight, and that they fit properly in their respective holes. It is abolutely necessary that each pivot should have *some* side-shake, but it is also very important to *guard against too much side-shake*, as such an excess causes loss of power and uncertainty in the action of the escapement. A desirable amount of end shake should be from ·02 to ·05 mm. As soon as these points have been found to be correct we are ready to try the "lock" and the "drop". In describing the pallet action, we made the clear statement that the lock should amount to from 1° to 1° 30⊠. This statement is, of course, of no practical use unless we are equipped with the necessary instruments for measuring this angle. We may, however, use the thickness of the pallet stones for comparison and obtain practically the same results, by making the amount of lock equal to 1/10 to 1/8 the thickness of the stone, from the locking to the let-off corner. This corresponds very closely to the above angular measurements. If the pallet stones are to be moved, in order to change the amount of lock, it is very important to first consider what will be the effect of a certain move, besides the alteration of the lock. The drop, for example, is effected very rapidly by moving the *L* stone. Hence if the drops are equal, we should make the change in the lock by moving the *R* stone. If the lock is too strong, and the drop is largest on the outside, the *L* stone should be moved. If the lock is too strong, and the drop is the largest on the inside, it is necessary to move both stones. Move the *L* stone out a small amount, and move the *R* stone in until the lock is correct. It is also well to recognise that the drop may be modified to a certain extent by moving the pallet stones, close to one or the other side, in the slots; as there is always some room allowed for the shellac which is used for holding the stones. The moving of the pallet stones in or out in the slots will also affect the draft feature of the escapement; this is a point which we should bear in mind whenever we make a change in the position of the pallet stones. The effect of moving the *R* stone out is to increase the draft on both stones, whereas if the *L* stone is moved out and the *R* stone in, it will decrease the draft. In order to ascertain that the escape wheel is correct, the lock and the drop should be tried with every tooth in the wheel on both pallet stones. This should be done with the bankings adjusted close, so as to just permit the teeth to drop. And the best way to try this, is to move the balance slowly with the finger while the pallet action is observed through the peepholes. After completing the adjustment of the pallet action the jewel pin action is next to be considered. The fork should swing an equal distance to each side of the centre line when the pallet is banked to drop. If we find that it moves farther on one side than the other, it will be necessary to bend the fork close to the pallet a sufficient amount to bring it in line. This is called "adjusting the let-off". The test for the let-off is to see that when the pallet is banked to drop, the jewel pin is just as close to the corner of the fork, in passing out, on one side as on the other. This test is correct, provided that the fork is of equal length on both sides of the slot, as it should be. The test for the fork length is that it should allow the jewel pin to pass out on both sides when the pallet is banked to drop. This is the maximum length which is allowed for the fork.

The test for short fork is to move the balance so as to unlock the pallet, then reverse the motion and see that the pallet is carried back safely to lock by the jewel pin. This should be tried on both pallet stones. It is, however, customary to try the shake of the fork when the centre of the jewel pin is opposite the corner of the fork, and not to allow the pallet to unlock from this shake. In order to ensure perfect freedom in the jewel pin action, the jewel pin should be from ·01 to ·015 mm. smaller than the slot in the fork. The safety action is also adjusted, while the escapement is banked to drop. The guard pin should be made just barely free from the roller when the fork is against the banking, and this should be tried carefully on both sides. If this is done correctly, the roller will have the necessary clearance when the bankings are opened to allow for the slide.

The operation of moving a pallet stone is one that requires a great deal of experience before one is able to do it satisfactorily except by repeated trials. Special tools called "pallet warmers" have been devised for holding the pallet during this operation. In the simplest form this tool consists of a small metal plate, about as large as a 12-size barrel, with a wire handle by which it is held while it is heated. This plate should have one or more holes drilled in it as clearance for the pallet arbor. An improved form of this tool is shown in Fig. 78A. The pallet is placed top-side down against this plate, and the whole of it is warmed over the alcohol flame until the shellac is softened so the stones can be moved.

A good way of applying shellac for the fastening of pallet stones is to warm some stick or button shellac, over a flame and pull it out in long threads of about .5 mm. diameter. Shellac in this form is very convenient to use, as it is only necessary when the pallet is heated to the proper temperature, to touch the end of this thread to it at the place where the shellac is wanted. With a little practice one can learn to deposit just the right amount. After the pallet is cold all shellac on the surface should be cleaned off carefully with a scraper made of brass or nickel.

The Jewel Pin Action.—The fork and jewel pin action involves two distinct functions; the impulse and the unlocking. In order to illustrate and make this statement clear, we will consider the different parts of the escapement in a normal position, Fig. 75. The hairspring, controlling the balance, has brought the fork, by means of the jewel pin, to the normal position of rest.

This leaves the pallet in a position where the impulse face of an escape wheel tooth will engage the impulse face of one or the other of the pallet stones, in this instance the *R* stone. Assuming the parts to be in this relation to each other, it is evident that when power is applied to the escape wheel, the escape wheel tooth which is engaging the *R* stone will cause the pallet to turn on its pivots, and this impulse is transmitted to the balance by the fork acting on the jewel pin. The impulse being completed, the escape tooth drops off from the *R* stone, and the second tooth forward comes to lock on the *L* stone, with the fork resting against the banking, as shown on Fig. 76. The fork slot is now in such a position that the jewel pin may pass out perfectly free, and this condition is necessary because the impulse which was given to the balance imparted to that member a certain momentum, causing it to continue to turn in that direction until this momentum is overcome by the tension of the hairspring. During this part of the motion, which takes place after the impulse, the jewel pin leaves the fork entirely, but the instant that the momentum in the balance is overcome by the tension in the spring, the balance will start to turn in the opposite direction, the tendency of the spring being to bring the jewel pin to the centre line. Before reaching this point, however, the jewel pin has to perform the very important function of unlocking. At the completion of the impulse we left the fork resting on the banking, with the fork slot in such position that the jewel pin *passed out* perfectly freely, and, figuring on the assistance of the draft and safety action, which will be explained later, we are justified in expecting that the jewel pin shall *pass in* to the fork slot perfectly freely. The instant the jewel pin has entered the slot, and comes in contact with the fork, the work of unlocking begins. And here is to be noticed that for every tick of the watch, the pallet and fork is started from the condition of rest, by a sudden blow of the jewel pin. And not only the pallet is started, but the *whole train has to be started in the reverse direction*, against the power of the mainspring, to unlock the escape wheel in order to receive another impulse. The jewel pin passing out on an excursion, the same as on the other side, returns to unlock, receives a new impulse, and so on, at the rate of 18,000 times per hour. In view of the above it is evident that lightness, as far as it is consistent with strength and wearing quality, is an essential feature in the construction of the several parts. It was once considered necessary to attach a counterweight to the pallet in order to get it in poise, but with the modern light construction of pallet and fork, it has been proven beyond a doubt that the ordinary form of counterpoise was worse than useless, inasmuch as it involved an added mass of metal whose inertia must be overcome at each vibration of the balance.

The Safety Action.—The function of the safety action is to guard the escapement against unlocking from sudden shocks, or outside influences, while the jewel pin is out of engagement with the fork. In the lower grades of watch movements this guard duty is assigned to the edge of the table roller and the guard pin. The passing hollow, a small cut in the edge of the roller, directly outside the jewel pin, allows the guard pin to pass the centre line during the jewel pin action. This form of safety action is called "single roller" and is shown in plan in Figs. 77 and 78. As will be seen from this drawing, the edge of the roller is made straight, or cylindrical, and the guard pin is bent in such a way as to present a curved portion to the edge of the roller. The advantage gained from this construction is that the guard pin can be adjusted forward or back by simply bending it at the base, without its action being in any way affected by a reasonable amount of endwise movement of either the balance staff or the pallet arbor. The double roller escapement, Fig. 76, presents a more desirable form of safety action, for two reasons: first, the intersection of the guard pin with the roller is much greater, making it perfectly safe against catching or wedging; second, any shock, or jar, causing the guard pin to touch the roller, will have less effect on the running of the watch, because the impinging takes place on a smaller diameter. The diagrams, Figs. 77 and 78, illustrate the above statements. The wedge action of the guard pin, when it is brought to the roller, is represented by the lines *C* and *D*, which are at right-angles to the lines *A* and *B*, thus forming tangents to the points of contact. It will be seen that with the single roller this wedge is 17°, whereas in the double roller it is 30°, a very considerable difference in favour of the double roller.

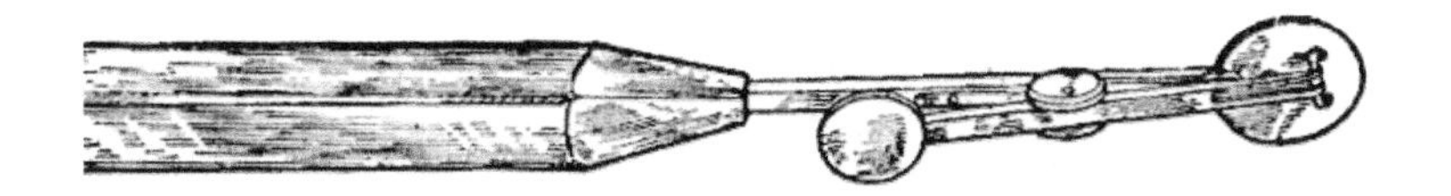

FIG. 78A.—A PALLET WARMER, FOR SETTING THE PALLET STONES,

Directions for Putting the Escapement in Beat.—An escapement is said to be in beat when it requires the same amount of power to start the balance in one direction as in the other. This should be tried with the mainspring only partly wound up, by arresting the motion of the balance with a pointed object held between the heads of two balance screws, and allowing the balance to move slowly, first in one direction and then in the other. If it appears to require more power in order to let off on one side than the other, it is said to be "out of beat", and it should be corrected by turning the hairspring collet a certain amount, on the balance staff, until it takes the same amount of power to let off on one pallet stone as on the other. This is usually done without removing the balance, by reaching in over the top side of the hairspring with a special tool made of small steel wire and flattened at the end so as to enter the slot in the collet. Great care should, however, be exercised in doing this work, so as to avoid bending the hairspring out of true.

CLEANING WATCHES.

Watches require cleaning for two reasons. They may be dirty through dust having entered and settled upon the moving parts, or the oil may have dried up around the pivots and become sticky. Thus a watch may want cleaning although it is not really dirty at all and may not even have been worn.

To cleanse dirt and sticky oil from watches, they must be taken apart and the pieces put in benzine or petrol, both of which are solvents of grease of all kinds. They are then brushed clean with a watch brush charged with a little dry chalk.

Key-wind Geneva Watches.—To take a watch apart, the first thing is to take it out of its case. Watches are fastened in their cases in many different ways. Taking key-wind Geneva watches first, they are generally held in by a short pin or pins on one side, just entering the edge of the case, and by a dog screw, sometimes two dog screws, on the other side. The dog screws can be seen when the case back is opened, and may be on the top plate, overlapping the case edge a little, or they may be on the bottom plate. Fig. 62 shows a dog screw. Generally they are cut away as shown, so that a half-turn enables the movement to be taken out of its case; but sometimes they are not so cut, and have to be entirely removed. In replacing a movement fastened in like this, the pins are first got into position, and then the movement is pressed down to its seating and the dog screws turned. All watches, without exception, are put in from the case front.

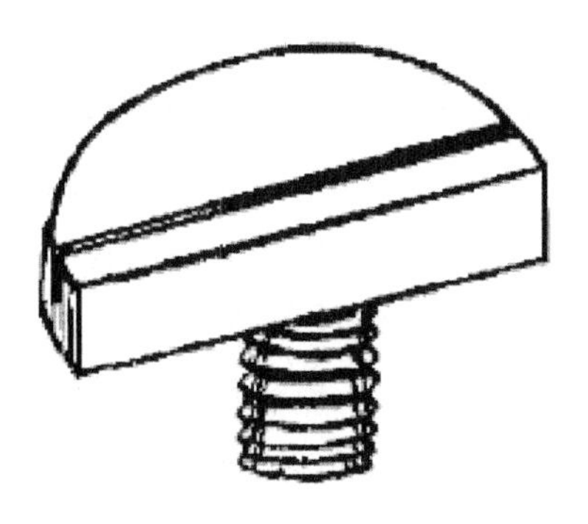

FIG. 62.—DOG SCREW.

When out of its case, remove the hands by drawing them off with cutting nippers; when there is no room to insert the edges between the minute hand and the hour hand, insert the edges under the hour hand and draw both off together. Draw the seconds hand off in the same manner. Special pliers and tweezers are made for getting hands off, and can be used by those whose stock of tools is not limited; they are safer and easier to use than cutting nippers. Using a pocket-knife to lever them off is the worst method, and should not be practised.

Remove the dial. If a white enamel dial, it may be fastened by pins through its feet, in which case they are removed with the aid of the small blade of a pocket-knife; or its feet may be slotted and held by two dog screws, which require half a turn *in*, and the dial can be lifted off. Do not force an enamel dial; it will not spring, but crack. Gold or silver dials, and some white ones, are snapped on like box lids over the top edge of the watch plate, and can be removed by levering up at the edge with a pocket-knife.

The dial off, the motion wheels will be seen. The hour wheel which carries the hour hand can be lifted off; the minute wheel can also be lifted off its stud. The cannon pinion is pushed friction tight on to the centre arbor, which carries the minute hand and passes through the hollow centre pinion.

If the arbor projects through the cannon pinion, hold the watch in the hand, and give a smart tap with the hammer and it will probably go through. This will at least loosen it so that by grasping the square at the back in a pair of cutting nippers and the cannon pinion "pipe" in a pair of brass-nosed pliers it can be twisted off. If the arbor end is flush with the cannon pinion top, twist it off as above described; or if too tight for that, lay the watch over a hole in the graduated stake, and with a small hard steel, punch tap the arbor through.

If the watch is a horizontal, that is, having a cylinder escapement, before the balance is removed the mainspring must be let down. This is easily done by placing a key on the winding square and holding the click back, letting the key run slowly back in the fingers. In 3/4-plate watches the click is covered by a cap fastened by three screws, which must be first removed; the balance-cock screw can then be removed and the cock levered up. It will come away together with the balance and hairspring. A little shaking will free it from the scape-wheel teeth. Unscrew the scape cock and remove the scape wheel. Take off the top plate and remove the remaining wheels. In a bar movement each wheel is held by its separate bar or cock, and care should be taken not to mix the screws. In a lever watch the balance may be removed before letting the mainspring down, but this should be done before removing the pallets and scape wheel.

The benzine is kept in a glass jar with an airtight cover, and must be kept well away from a flame. Into this jar place the plates and wheels, all except the barrel and balance. Let them soak while the barrel and mainspring are taken in hand.

The barrel of a 3/4 plate is the simplest. Underneath will be found the "stopwork." Hold the winding square in a pair of sliding tongs and prise off the centre steel piece or stop finger. This may be only pushed on friction tight or may be pinned through. Remove from the tongs, and with a screwdriver prise off the barrel cover. This is snapped in a groove in the barrel edge. Then remove the steel arbor. Put arbor and cover in the benzine. Do not pull the spring out of the barrel if it looks all right, but wipe it out well with dry tissue paper, free from oil and dirt. Sharpen a watch peg and clean out the centre hole, twisting the peg round and scraping it clean again until the hole no longer marks the peg. Take the arbor and cover out of the benzine, and, holding them in tissue paper, brush them clean and dry. Peg out the hole in barrel cover.

For re-oiling the barrel and mainspring, a bottle of the best French clock oil should be used. Having brushed the barrel clean and its teeth also, take the arbor and oil its top pivot, where it works in the barrel, with the clock oil. Place the arbor in position and turn it round, seeing that it is hooked properly in the eye of the spring. Apply oil to its bottom pivot, where the cover comes, and some oil to the coils of the spring. Then snap the cover on with the fingers or by pressing it against the wooden edge of the work board. Never use pliers for this. The cover goes on in one position only. There may be a small projecting pin in the barrel groove that goes in a small notch in the cover edge; there may be a dot half on the cover edge and half on the barrel edge; there may be a dot on the cover near the edge and another on the side of the barrel to match it; or the slot for removing the cover may merely have to go next the dot on the barrel side.

When together, hold the square in the sliding tongs and the barrel in the fingers, and feel if the arbor has "endshake" and is quite free. Then wind the spring up to the top and count its turns. As it runs back slowly, feel if it catches or binds inside the barrel. Suppose the spring gives five turns. The stopwork allows four to be used. There will be a turn to spare, so divide this between the top and the bottom. "Set up" the spring half a turn. To do this, wind it up half a turn and hold it there while the stop finger is replaced in the position shown in Fig. 63, seeing that the star wheel A is as shown in the figure. Upon releasing the barrel, the stopwork will then hold the spring wound up half a turn. Wind up the spring to the top to see that all is right. There should be half a turn of spring to spare, thus preventing the spring tearing at its hook in the barrel.

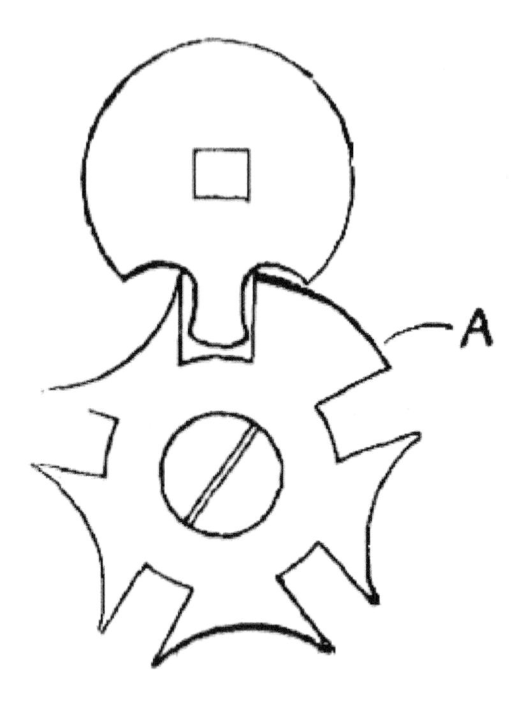

FIG. 63.—GENEVA STOPWORK.

The barrel of a "bar" movement is more complicated. First take off the top cap or "chapeau" that covers the arbor. Remove the stopwork and barrel cover. Take hold of the centre coil of the mainspring with fine pliers and ease it off its hook. Then pull it out of the barrel. Place a key (birch, black-handled universal keys are always used by watchmakers) on the square, and grasp the arbor inside the barrel with pliers and unscrew it. The barrel, arbor, and bar will then come apart. Put all in the benzine except the mainspring, and clean and peg out. Wipe the mainspring with tissue paper free from grease. To put together, place the arbor in the key and oil where the bar comes and where the barrel turns. Put on the bar and the barrel. Screw on the arbor inside. Place the eye of the mainspring upon the arbor and hook it, and, holding and guiding it with the fingers, wind it in the barrel with the key. When in it will generally hook itself. If not, press the spring well down and wind it up a turn or two. This will hook it. Oil it and place on cover and stopwork as before described. Take out the wheels from the benzine, and, holding them in tissue paper, brush them clean and dry. See that the teeth are clear, and peg out the pinion leaves clean last of all. Clean the cocks, bars, and plates in the same manner.

The watch brush should be soft and thick, and just lightly rubbed on a billiard chalk before cleaning each piece. The chalk is to keep the brush clean as much as to clean the parts. The tissue paper keeps the fingers from soiling the parts and keeps the brush from rubbing the fingers. The constant contact with the tissue paper cleans the brush.

Peg out the pivot holes very carefully with a fine peg point, scraped thin. If a peg point breaks in a pivot hole, sharpen a fine-pointed peg to a somewhat blunt angle, so as to be both sharp and strong. With this push out the broken peg from the back. Jewel holes covered by "endstones" must have the endstones removed and rubbed clean on a piece of wash-leather. The holes must be pegged and the endstones replaced. When all are cleaned, screw the bottom plate, or "pillar plate," in a watch holder (Fig. 64). This avoids handling the clean plate, and leaves both hands at liberty for working upon it.

FIG. 64.—WATCH MOVEMENT HOLDER.

If a 3/4-plate watch, put the third and fourth wheels in place, then put *watch oil* on the centre wheel and barrel pivots and put them in place. Put on the top plate. Put a piece of tissue paper over it and press lightly on it with the fingers, slipping the top pivots of the wheels into position with a fine pair of tweezers. Use no force in this operation. When right the plate will drop down in position.

Watch oil is not taken direct from the bottle for oiling watches. Such a proceeding would soon render the oil useless.

An oil "cup," with a cover, is used for the workboard. Into the cup about two drops of oil are placed, by dipping the oiler into the stock bottle and transferring the drop into the cup. To apply oil to a pivot, the oiler blade is dipped into the cup and a minute quantity of oil transferred to the pivot. Two drops in the cup should be sufficient for a day's work.

If a bar watch, screw on the barrel bar; then put in the centre wheel and bar; then the third and fourth wheels and their cocks.

Put in the scape wheel and pallets if a lever, placing just a little oil on each pallet face before putting them in.

Each wheel should be tried carefully, to see if it has enashake, as the watch is put together. Every wheel and arbor in a watch *must* have endshake, or they will bind and stop the watch. The amount of shake or lift must be enough to see with an eye-glass, but should not be excessive. An amount of shake equal to half a pivot length is far too much. Fig. 65 shows a fair amount in proportion to a pivot. When wheels are held by cocks, the endshakes can be regulated by inserting small pieces of tissue paper, or thicker paper if necessary, under either the front or back end of the cock. More endshake can be given to wheels under a 3/4 plate by putting a paper washer upon the pillar top under the plate. The barrel arbor alone need have no endshake between the plates.

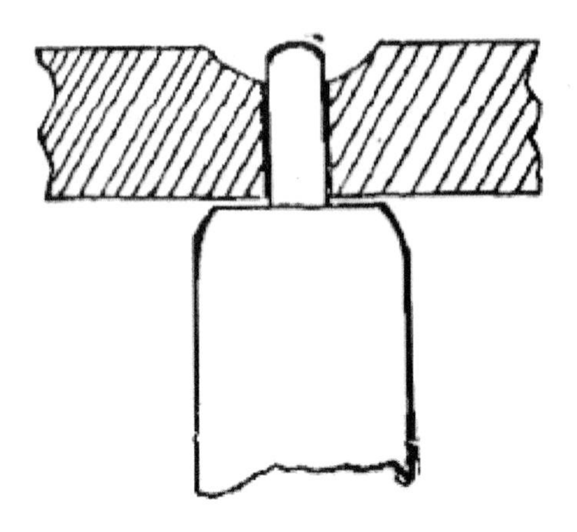

FIG. 65.—A CORRECT ENDSHAKE.

The set-hands arbor can be cleaned and put in, and the cannon pinion pushed on home. This may require a tap with the hammer and a suitable hollow punch, passing over the cannon pinion pipe and resting on the leaves.

The balance and balance cock, etc., can now be taken in hand. The hairspring stud must be first removed from the cock. Usually the stud is pushed friction tight into a hole in the cock. In this case the cock edge may be rested on a stake and the stud pushed through from the top. When loosened thus, complete its removal with tweezers from underneath. The cock can then be taken apart and cleaned with benzine, the jewel hole and endstone cleaned and replaced, together with the regulator index. The hairspring should not be removed from the balance, but all may be dipped into the benzine. It can be dabbed, hairspring downwards, on paper to remove the benzine, and given a minute to dry off. The cylinder should be cleaned out inside with a peg. The pivots and balance rim can be cleaned by pith.

Place a small quantity of watch oil in the jewel hole, replace the hairspring stud, pressing it well down in place, and see that the hairspring goes well between the curb pins in the regulator without pressing hard against either of them. Put a little oil on each scape-wheel tooth and in the lower balance Jewel hole and put balance in. Screw the cock down very carefully, seeing that the pivots go into the holes and that the balance is free before screwing quite tight. The endshake of the balance is very particular. A little, but only just perceptible, is wanted. It can be regulated with tissue paper, as before described. The watch can then be wound, and oil can be applied to the top and bottom pivots of the train wheels and scape wheel. The minute wheel and hour wheel can be put on (these need no oil) and the dial replaced. If a dial held by dog screws, draw the dog screws outwards to tighten, so as to draw *up* the feet; this prevents the dial rattling. When the dial is on, see, that the hour wheel has a little endshake, or lift, under the dial. If too much, put paper collets over it under the dial. Put on the seconds hand, seeing that it lies close to the dial and flat, but does not touch it anywhere. Put on the hour hand, seeing that it has a little shake and is not bound against the dial, also that it just clears the seconds hand. Put on the minute hand, and, resting the set-hand square on a steel stake, tap it on with the hammer. See that the minute hand is well clear of the hour hand, but that it does not stick up enough to touch the glass.

Dust out the case and replace the movement.

The various parts of this movement have not been illustrated, as the figures and accompanying descriptions in the introductory chapter will be found sufficient.

Particular care must be taken that oil is applied to every pivot. Neglect of this leads to certain disaster. The friction causes rapid wear and rust. The rust powder, in itself a polishing and cutting powder, rapidly cuts the pivot away, until it disappears entirely, leaving only a mass of rust in the pivot hole.

Keyless Watches.—This description of the method of cleaning and putting a Geneva watch together applies to key-wind and keyless watches alike as far as it goes. But keyless watches require some further attention as well. The first difficulty with a keyless Geneva watch is to get it out of its case. They are fastened in in various ways. Dog screws like those shown in Fig. 62 actually hold the movement; but before it can be removed the winding stem, and sometimes the set-hand side push piece, must be taken out.

In some watches a small screw is found in the case pendant, and removing this enables the button and winding stem to be pulled straight out. In most a small steel screw either at A or B, Fig. 66, has to be withdrawn a turn or two before the stem can be drawn out. If there is also a screw at C opposite the side push piece, draw it out and remove the push piece; but in most watches the movements will come out without removing this, and no screw will be found at C.

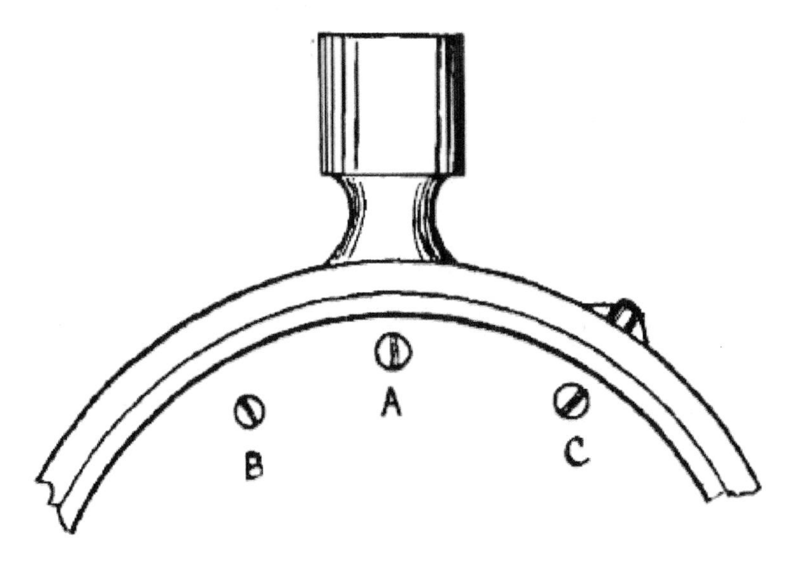

FIG. 66.—POSITION OF PENDANT SCREWS.

Occasionally the screw fastening the winding stem in is underneath the dial, and the hands and dial must be first removed to get at it. A very few watches (mostly by "Patek Phillippe") have the winding stem fastened in such a way that to get the watch out of its case the bar holding the winding wheels in the watch must be taken out and several other parts also, as well as the hands and dial, to remove it.

A keyless watch should have the mainspring "let down" before it is taken out of its case. This is done by holding the winding button in the fingers, while the click is held back, and allowing it to slowly run back.

There are many kinds of keyless work, and these will be described in a chapter to themselves. In cleaning the watch, all keyless wheels and parts should be taken off and cleaned in the benzine. In putting together, all want well oiling where they rub against each other or the watch plates or bars. In taking them apart, it is well to remember that top keyless winding wheels are frequently fixed by a large central *left-handed* screw, which, of course, requires turning to the right to unscrew it.

English Watches.—English watches may be full plate or 3/4 plate, and keyless or key-wind.

Taking full-plate, key-wind fusee watches first, the movements are generally fastened in their cases by means of the joint pin at the Fig. 12. This should be removed by a "joint pusher." It should always be used held in the hand when possible. When enough pressure cannot be got in this way, it may be screwed fairly tight in the vice, as shown in Fig 67; the watch is held up to its point, and a smart blow given to its head with the hammer. This will dislodge the most stubborn joint pin. Before resorting to this method the balance had better be removed, for fear of injury to its pivots.

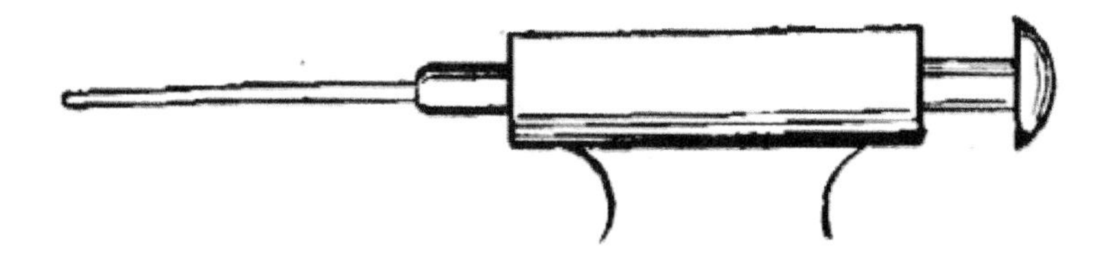

FIG. 67.—USING THE JOINT PUSHER.

When out of its case, remove the hands, take out the balance cock and balance, and remove the dial. Dial pins may be removed by the aid of a pocket-knife, levering them forward. Sometimes a pin cannot be got out in this way because none of it projects. In such a case it must be got out by applying pressure from the back, or inner end, by inserting a screwdriver blade and pushing, or in stubborn cases using a steel draw-hook to pull them forward. Such a hook is shown in Fig. 68. It is inserted behind the inner end of the pin and pulled forward.

FIG. 68.—PIN DRAW HOOK.

The mainspring must then be let down. It is held by a ratchet and click under the pillar plate. Loosen the click screw half a turn, place a key upon the square, and let it down. Sometimes there is not enough square for a key to hold. Then screw a pin vice on to the fusee winding square, screw the pin vice in the board vice, like Fig. 69, and, holding the movement firmly in the hand, take off the under bar that holds the third and fourth wheels. Take out the third wheel, replace the bar and screw it down, then let the movement slowly turn round in the hand as the mainspring unwinds itself.

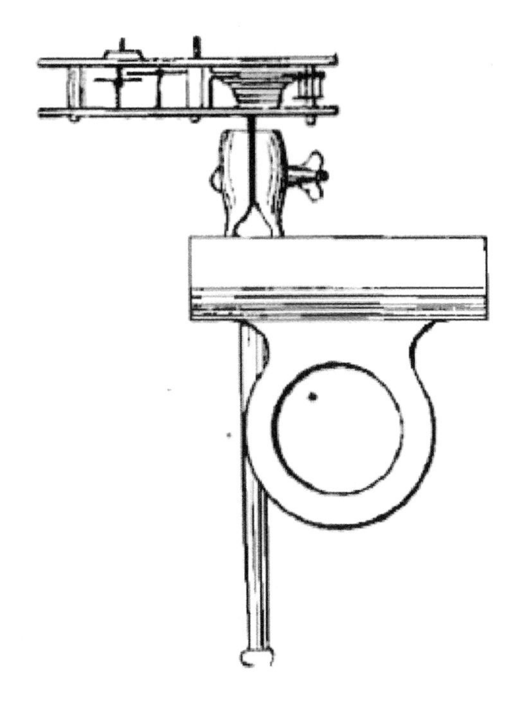

FIG. 69.—LETTING DOWN A MAINSPRING.

When let down, take off the barrel bar or name bar and remove the barrel, having unhooked the chain first. Then take out the four pillar pins that hold the top plate, and gently raise the plate a little to enable the lever to be lifted out of its pivot holes and withdrawn. If the plate is taken off before taking this precaution, the lower pivot of the pallet staff will be bent or broken, as the lever will be caught by the potance. The potance is the brass cock screwed to the under side of the top plate to carry the lower pivot of the balance.

Then lift off the top plate and remove the fusee and wheels, etc. The cannon pinion can be removed from the centre arbor by grasping its square in a pair of cutting nippers and pulling or twisting, holding the centre wheel tight in the mean time.

Plates, wheels, bars, etc., can be put in benzine. The barrel and mainspring can be cleaned as in a Geneva watch. The chain can be wiped in slightly oiled tissue paper. The fusee, if the clickwork inside it seems sound, need not be taken apart, but merely brushed clean, and *not* put in benzine. If the fusee winds too hard, run a little oil between the edge of the steel maintaining ratchet and the brass fusee body.

In putting together, put the plate in the holder, place the third wheel in position, oil the bottom pivot of the centre wheel and place that in position, oil the fusee pivots top and bottom, and put the fusee in. Put the maintaining detent in place the wrong way round, as it then stands up more easily and can be turned right afterwards. Put in the fourth and scape wheels; then put on the top plate, and before it is got down to its place, introduce the lever and get it in position with tweezers; then get the pivots in their holes and the top plate down, pinning it on. The barrel goes in next and its bar, then the. cannon pinion. The maintaining detent (whose point should have been filed up sharp before putting in) can be turned round, and should be examined to see that it engages properly. The chain now has to be put on.

Turn the fusee and barrel round so that the chain holes are outward. Hold the movement vertically so that the chain can be dropped clear through it from fusee to barrel. Then hold the movement in the left hand, and with tweezers insert the barrel hook in its place. Place the thumb of the left hand upon it to keep it in place, and putting a key on the barrel square, wind the chain upon it until it is nearly all on. Hook the other end in the fusee and turn the barrel round a little further to draw the chain tight. Now place on the barrel ratchet and "set up" the spring half a turn, or three-quarters if the spring will allow, and screw the click tight.

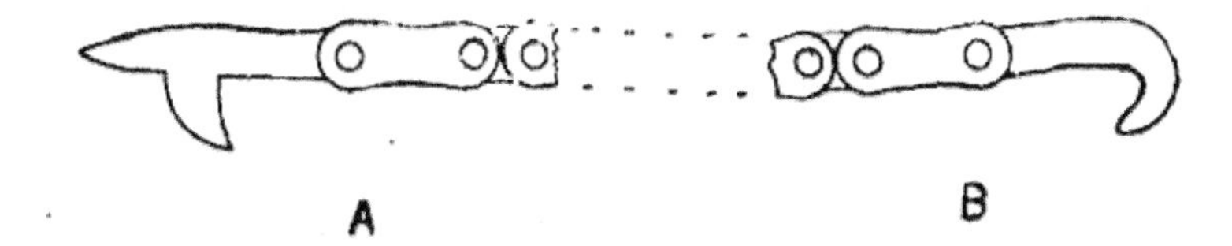

FIG. 70—CHAIN HOOKS.

In setting up the spring the want of a third hand is often felt to put the click into the ratchet teeth. This want can be supplied by holding a long peg in the mouth and using it to manipulate the click. Fig. 4, p. 3, shows the arrangement of a barrel, chain, and fusee. Fig. 70 shows the two ends of a fusee chain. A is the barrel hook, B is the fusee hook.

When all is wound upon the barrel and set up, the watch can be carefully wound up. As this is done, see the chain runs straight to the fusee and does not drag aslant, or it will get out of its groove.

Now clean, the balance and cock, oil the balance holes, the top pivot holes of the centre, third, fourth, and scape wheels, and lever (a very little indeed to the lever). If much is put to the lever it may run between the lever and plate and clog. Put in the balance. In a watch in which the hairspring is above the balance (oversprung) the stud will simply be screwed in place. In an undersprung watch the spring will require re-pinning in its stud, which is a fixture on the plate. And it must be pinned in at exactly the right place to set the watch "in beat." To test for this, wedge the fourth wheel with a peg or a pivot broach, and if in beat, when the balance is at rest the ruby pin in the roller will be in the lever notch, and the lever will stand midway between its banking pins. Adjust the hairspring until this is so, and then see that it passes between the index or regulator "curb" pins centrally and plays between them evenly, especially when the regulator is at "slow." See also that it lies flat and does not touch the balance or the plate.

Apply a little oil to the points of the scape-wheel teeth, oil the bottom pivots, put on the motion work, dial and hands, and put in the case again. See particularly that the dial pins fit well and go in tight; also that none of them touch the wheels, etc.

The fusee is out of date in lever watches, and not now made except in special cases. Most modern English full-plate watches and all American ones have "going barrels" like Geneva watches.

They are generally arranged so that the barrel bar can be taken off and the barrel removed without taking off the top plate. They all should be so made. The mainsprings of these watches can be let down by a key on the winding square, while the click is held back. In putting them together, it is advisable to put the barrel in place with the rest of the wheel-work before putting on the top plate. The barrel then steadies the plate and helps to hold it in the right position while getting the lever in. Otherwise these watches are cleaned in the same manner as fusee watches.

English and American 3/4 plate key-wind watches will present no special difficulties. The principal difference is that these watches have the top plate cut away; and the balance and escapement are held by separate cocks screwed to the pillar plate. This has several advantages. First, it secures a flatter watch; second, the balance is less likely to be crushed and injured through accident; third, the hands can be set from the back; fourth, the escapement can be removed separately.

Besides these, other forms of English full-plate watches will be met with—the verge, the cylinder, and the duplex more especially, and a few pocket chronometers.

Verge Watches, the oldest form of all, now not made, are rapidly dying out. Still, it may be many years before they are all worn out or disused, and the country watchmaker still has them with him.

A verge resembles a lever up to the fourth wheel. The fourth wheel is a "crown wheel," and drives the scape pinion and wheel, which is carried under the top plate by a brass "follower," and the potance, as in Fig. 71. A is the potance, B the "follower," and C the scape wheel. The verge or axis of the balance generally runs in brass holes top and bottom.

In cleaning these watches, do not remove the follower to take out the scape wheel, but unscrew the potance. This is less likely to disturb the escapement. The dead brass holes in follower, potance, and balance cock must be very carefully cleaned by pegging out, and care taken not to break a peg in. In putting together, apply oil to the lower verge hole first and to the scape pivot hole in the potance, as these cannot be got at afterwards. Then put the scape wheel in position and oil the follower hole. The train wheels and top plate can then be put on. Put on the chain before putting the verge in. Set in beat by seeing that it cannot be stopped on either pallet, but starts off immediately it is released. If the balance cock has to be removed for any purpose after the watch is wound up, the fourth wheel must be wedged as a precaution. If not, the train may run and damage the scape-wheel teeth. Put no oil on the verge pallets.

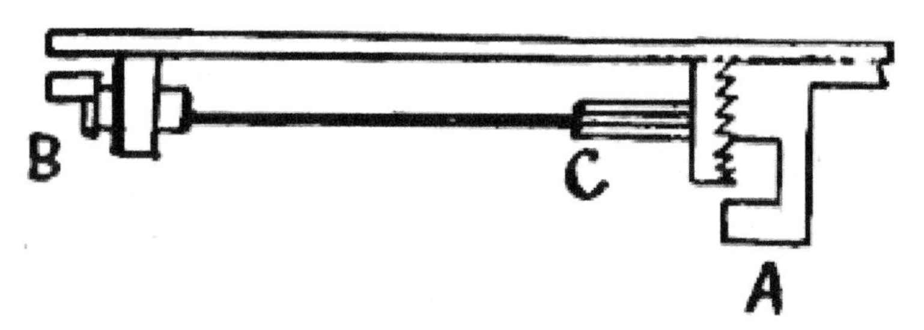

FIG. 71.—VERGE SCAPE WHEEL.

English Cylinder and Duplex Watches.—The same remark applies to these watches. The balance cocks must not be loosened while there is any power on the scape wheels. Otherwise these watches are cleaned just as levers. Cylinders require oil inside them and on the scape-wheel teeth. Duplex watches require oil on the ruby roller and on the point of the long impulse pallet.

Pocket Chronometers require great care in handling. The balance should not be removed without first wedging the fourth wheel, or there is danger. The delicate detent and scape wheel should be put in a safe place while the rest of the watch is being cleaned. When putting the watch together, if a full plate, leave the detent out, as it can be put in last thing. To clean the detent, lay it flat on clean paper and hold down the foot with tissue paper. Brush it gently to remove dirt, and with a fine pointed peg clean the locking ruby and the point of contact of the gold spring and detent point.

A chronometer only requires oil to its pivots; none should be put on the scape-wheel teeth, the locking stone, the gold spring, or the roller pallets.

English Keyless Watches, whether full-plate or 3/4-plate, will come straight out of their cases without first removing the winding stems.

American Keyless Full-plate Watches generally require the removal of a small set-screw in the case pendant, and the button can be drawn off. This allows the movement to come out.

Three-quarter-plate American Watches are of many patterns. Some are made just as the full plates above described; others require the loosening of a small screw on the top plate, like some Swiss watches. Although they vary very much in pattern, none will present any special difficulty in taking out of their cases.

The keyless work of all requires cleansing in benzine, and plenty of oil applying to all points where friction occurs.

General Remarks on Cleaning Watches.—In taking a watch apart, everything should be carefully tried before removing, to discover faults. The neglect of this causes a great deal of wasted time. It is better to find a fault when taking apart, and remedy it before cleaning, than to find it when the watch is cleaned and put together. Thus, when taking off the hands, see if they have been touching the glass, or if they are too loose or bind against each other. Test the escapement. Try all endshakes, feel all depths, and look for broken or cracked jewel holes before anything is cleaned.

In putting together do not finger cleaned parts. Any spots of black or dirt on the plates, that will not brush off, remove with a peg point. If the gilding has got tarnished, brush it with wet benzine on the brush. If a nickel movement has become tarnished, remove it with spirit on tissue paper, or, in bad cases, with rouge on a cork. Always remember to put oil to jewel holes with endstones before the wheels are put in. Enamel dials can be cleaned with a damp cloth. Gold hands can be laid on a cork and rubbed gently with another cork and rouge powder.

Screws, though they may appear to be interchangeable, are generally not so, and means should be taken to replace them in their correct positions. Some screws are dotted on the heads; if so, they should be placed in the holes dotted to correspond. One of every pair of jewel screws has a dot; this goes in the hole nearest to the dot on the edge of the jewel setting. Pins are seldom interchangeable, and dial pins should be kept apart from pillar pins.

Beginners should be very careful indeed in handling balances and hairsprings. These are the most delicate, as well as the most important, parts of a watch. A very little will bend a hairspring, and the least pressure will damage a balance pivot. Let a balance down gently into its lower pivot hole by its own weight. When putting the cock on and screwing it down, set the balance vibrating, and let it continue to do so while the cock is tightened. Any nipping or pressure upon the pivots will at once show itself by the balance ceasing to vibrate, and attention will be attracted to it before damage is done.

Before putting in the balance of a watch, just see finally if the train is all free by touching the centre wheel or great wheel teeth. If a cylinder, verge, or duplex, the train will run, and its freedom can be judged. If a lever, a little *back* pressure on the fusee or barrel teeth should cause the scape teeth to trip back past the pallets. A lever that will not do this readily is faulty.

A dot upon the centre steel piece of an English rocking bar must be placed towards a corresponding dot on the movement edge, if there is one. If not, put it on with the dot towards the centre wheel. Similarly, an English barrel cover must be put on with the opening next to the dot on the barrel, or, if there is no dot, next to the chain hook hole.

A slight finger mark or smear upon a newly cleaned plate can be removed with clean-cut pith, a clean surface being cut each time the plate is rubbed.

When jewel holes in English or American watches have endstones to cover them, the jewel settings are fitted closely in a sink turned out in the plate, and held in by two small jewel screws. Such jewelling should always be taken out by withdrawing the screws and pushing the jewel hole and endstone out with a flat-ended watch peg. They are both cleaned by rubbing them with the finger on a piece of washleather and pegging out the hole.

American jewelling, and some in machine-made English lever watches, is often fitted so tight in the plate that great force is needed to push it out. In such a case take care that the pressure is applied to the brass settings and not to the jewels themselves. Such tight jewelling can be replaced by pressing in with a flat-cut peg, or by making a flat-ended punch out of pegwood and using the hammer. Some watchmakers keep a few ivory punches for this and similar purposes.

To avoid danger to the hairspring while brushing a roller or cylinder clean, push the roller or cylinder and lower part of the balance staff through tissue paper.

As tweezers wear, the inside edges of the points become smooth and rounded, causing pins and other small parts to "shoot" from them. The remedy is to pass a sharp fine file over their inside faces and thin them a little, bringing the edges up sharp again.

To handle a dial or pillar pin safely, pick it up from the board with tweezers, lay it on the back of the hand, and again take it up in the tweezers. This time the points will get a firmer hold, as the hand back is soft and allows the points to go further over the pin.

CAUSES OF STOPPAGE OF WATCHES.

Small Faults.—A watch may be cleaned carefully and well repaired, yet may, when put together, have a bad action, or stop in the workman's hands or under his very eyes. Such watches often give more trouble and waste more time than has been previously spent on cleaning and repairing them.

Beginners, after they have vainly looked all round such a watch, are apt to give it up as a "mystery." There are no mysteries in watchwork. A stopping watch has a fault somewhere, however small, and it must be found.

A workman who looks out for faults as he takes a watch apart is not troubled with these "stoppers" so often as the careless man. An escapement should always be examined and tried before taking apart; all wheels should be examined as to endshake and side play, marks of fouling each other or the plates looked for. As the watch is put together again, everything should be tested as it is put in.

Often the cause of stoppage is a dial pin fouling a train wheel, or a screw put in its wrong place, its point going too far through somewhere. Stoppages in the train are most easily found. Taking care not to start such a watch, with a needle try the scape wheel and see if it has "power" on it. If not, try the fourth, third, centre, and so on in succession, until the point is found at which the power disappears. This gives a clue, and a bent tooth, a bent pivot, a tight endshake, or some cause of fouling must be looked for in that wheel or pinion. A watch that stops once per minute probably has a fault in the fourth wheel or pinion. If once per hour, it may be the centre wheel or pinion or the motion work. If every fifteen beats, a damaged scape tooth or dirt in the scape pinion. When power appears to go off at the barrel itself, it may be that the stopwork jams. A little roughness on the centre stop finger of Geneva stopwork is liable to catch the points of the star wheel. A shallow depth between stop finger and star wheel will also cause it to jam. The mainspring may bind in the barrel, or a cap screw point may be too long and bind it.

If a fault is suspected in the train and nothing can be seen to account for it, run each wheel separately in its frame to see that it has perfect freedom; then each two wheels together, and try all depths.

A watch that seems to have plenty of power but has no action has a fault in the escapement. After trying the scape depth, the banking shakes, run, and endshakes, see that the lever and roller do not touch, that there is no oil between the lever and the plate, that the hairspring lies flat and true, free of the balance, the plate, and everything else. If no fault can be found there, run the balance in its pivot holes alone, with the roller and hairspring removed. Let it run slowly, and observe how it stops. A bristle from the brush may be found sticking in its rim, or in the plate, just touching it. The balance rim may touch the fusee chain in a 3/4 plate. Then put on the roller and let the watch go to half time, without the hairspring, and see if it acts freely and does not catch anywhere.

A pivot may not come quite through a jewel hole, or the hairspring collet may touch the balance cock.

Much can be learned by listening carefully to the beat in various positions. If the balance is foul of anything, a striking will be heard. If a Geneva horizontal scrapes in one position, it is probable that the scape wheel touches either the bottom or top of its passage in the cylinder.

A watch that stops in one position only has generally a fault in the escapement. If the pivots and jewel holes are quite right, it is a fault caused by the movement of the balance, scape wheel, or pallets, owing to their endshake or side play. Something that is quite free with the endshakes one way fouls when they are the other way. Excessive endshakes may be present.

Note if the action is equal dial up and dial down. Any falling off will indicate a fault. To observe the action dial up, a small piece of mirror laid upon the bench is extremely useful, as the watch can be held steadily over it and the action observed at leisure.

A lever watch that falls off in action when the balance leans towards the lever has probably a fault in the roller and lever depth. The ruby pin may be a round one and not enter the notch properly. A flatted pin will cure this. There may be not enough, or too much, banking shake, or a roughness on the roller edge.

Watches that go all right lying down and hanging up, but stop in the pocket, will often be found to be too shallow in the safety-pin action. The safety pin in the lever can jam against the roller edge. Also, in such a case, look to the pallet depth and see if it mis-locks on any teeth, as this causes the safety action to jam.

If a watch stops when just wound up, see to the maintaining work, also that it is perfectly in beat and that the safety action is correct.

In full-plate English watches the barrel may rise just a little above the plate and foul the under corner of the balance cock, or the balance rim, in certain positions. If the mainspring is of the American pattern and has a brace, the top pivot of the brace may project and foul the balance rim or scrape the barrel bar. The point of the guard pin in the lever may touch the under side of an undersprung index.

The chains of these watches may scrape the back of the potance, or the inside of the cap. The barrels of 3/4-plate going-barrel watches may just touch the case edge and cause binding. The cannon pinion may be too low, and its teeth may scrape the plate.

The dial may press upon and bind the minute wheel or some of the points of the lower pivots.

Hands should always be looked at to see if they touch the dial or the glass and are free in the dial holes.

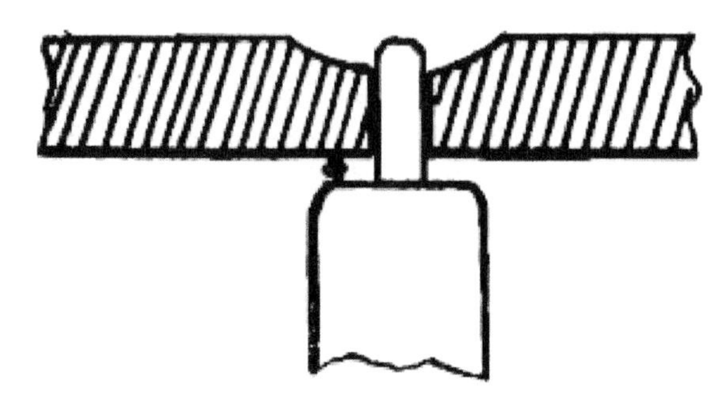

FIG. 181.—GRIT JAMMED BETWEEN PIVOT SHOULDER AND PLATE.

Some new machine-made English levers give a lot of trouble by stopping from the least dirt. It gets round the train-wheel pivots and jams them tight. This is caused by rough plates and rough pivots. Polish all the pivots and stone the plates round the pivot holes quite smooth. The gilding on the plates of these watches is like a honeycomb. The plates seem to be pickled to frost them deeply before gilding, presumably to cover up a want of smooth finish. The result is that grit is held by the surfaces of the plates and jams under the square pivot shoulders of the train wheels, as in Fig. 181.

If dial pins are not pushed in quite tight and do not fit well, they often work loose and come out. Being in the case, they rattle about until they get under one of the wheels and there stick. When a stopping watch is brought in minus one dial foot pin, look at once to see if it is not somewhere in the wheelwork. Similarly a small screw, if it works out, may jam the wheels.

A train wheel or pallet pivot that is short and does not come through its brass hole is apt, in an old watch, to wear a step in the hole. When much worn in this way, the step sometimes jams the pivot by stopping its endshake, as in Fig. 182. The remedy is to broach out and bush the hole.

In the holiday season, when people go to the seaside, *sand* is a frequent cause of watches stopping. A little gets in the pocket and finds its way into the watch case. One grain is sufficient to stop it, though, as a rule, many grains find their way in. A single grain fixed between two wheel teeth as a rule causes the stoppage.

Duplex watches and pocket chronometers, even when set in beat as carefully as possible, will occasionally stop in the pocket. Being single-beat escapements, it only needs a movement that brings the balance to rest at the right moment to stop them.

Verge conversions that have no maintaining work will also stop during winding sometimes. These should always be very carefully set in beat, to avoid this trouble as far as possible.

Watches with heavy escapements are also liable to stop when the hands are put back, especially if they move stiffly. For this reason, pocket chronometers should always have the hand work left easy, just tight enough for them to carry with certainty, but no tighter.

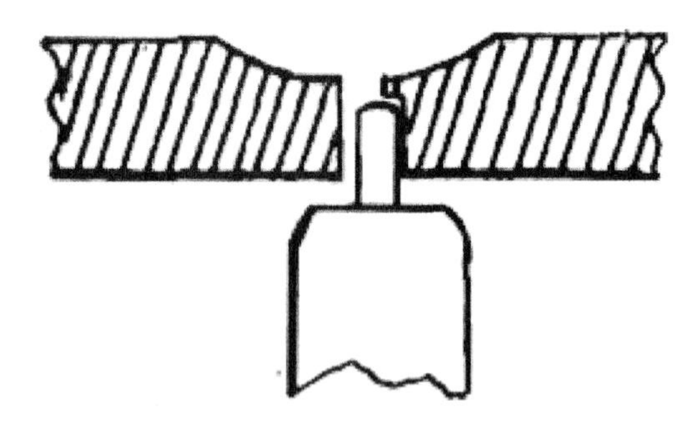

FIG. 182.—WORN PIVOT HOLE.

In some old English 3/4 plates the lever lies much too close to the pallet cock—so close that the oil applied to the top pallet pivot is almost sure to be drawn between the lever and the cock. The best way to serve them is to file away the under side of the cock each side of the pivot hole, and, if possible, turn a cone hollow around the top pivot in the body of the pallets.

Magnetized Watches.—In these times, when there is so much electrical machinery about, watches are often found to be magnetized.

If all the steel parts should be very strongly magnetic, the watch will not go at all, the balance being attracted by the screws, etc., around it, and the hairspring adhering to the balance. When only slightly magnetized, the watch will go, but the timekeeping will be very erratic. A watch becomes magnetized through being brought into the field of a powerful electro-magnet, such as a running dynamo or motor. To detect it, place a small charm or pocket compass flat over the balance cock, with the centre of the compass corresponding with the centre of the balance. If magnetic, the compass needle will vibrate as the balance does, or perhaps fly round and round. If not magnetic, the needle will be quiet. A perfect protection to a watch is an iron box, or what is commonly known as a "tin" box. It follows from this that ordinary watches, if in "gun-metal" cases (oxidized or blacked iron), and full hunters are perfectly protected from magnetic influence.

To demagnetize a watch, it should be revolved rapidly and brought into the field of a powerful magnet—say a dynamo—and gradually withdrawn. One way of effecting this is to first wedge the balance with tissue paper; then fasten a string to the bow and twist it up tight; suspend the watch by the twisted string close to a running dynamo, and let it untwist rapidly. As it does so, withdraw it out of reach. This is only a rough way. If a magnetized watch is sent to an electrician, he will demagnetize it more thoroughly for a fee of about 2*s*. 6*d*.

Careful tests show that once a watch is magnetized it is never *quite* free again. A fine watch with a close rate is for ever spoiled. The only perfect cure is to heat all the steel parts to redness, re-harden, and polish them again. This involves fitting a new balance, hairspring, mainspring, etc, and in most watches is not worth doing.

Karrusel Watches.—These watches have a few faults of their own that often stop them. The revolving carriage carrying the escapement must be dry, clean, and quite free. Where it bears and rubs on the plate, both top and bottom, must be channelled out by turning, letting it bear on circles only. These must also have no gilding on them, being stoned off smooth where the friction comes. Gilding wears off and works up into a black powder and causes choking.

The edge of the carriage is sometimes found to foul the end of the third wheel cock and the tops of the barrel teeth. The third cock is easily filed to free it. To free the barrel teeth, the barrel is best put in a step chuck and the teeth-tops bevelled down. The top third pivot hole is also rather liable to be left dry, and must be oiled before putting in the third wheel.

Bad Mainsprings.—A watch that has a poor action, the balance vibrating less than one complete turn, is always liable to stop from the least cause. Often such a watch will be found to have a poor, cramped up mainspring, and can be greatly improved by changing it for a good quality *lively* spring of the same thickness. Nothing pays better in watch materials than to keep a good quality of mainspring. A spring costing 4*s*. 6*d*. or 6*s*. per dozen, instead of 2*s*. 6*d*., will pay for itself many times over in making poor watches go well.

Unequal Rates.—Some watches after cleaning are round to go much slower hanging up than when lying. Generally this indicates a balance out of poise, or a hairspring not true in the centre. In very small ladies' horizontal watches this fault is often troublesome, and may be sometimes cured by deepening the cylinder depth, *i.e.* setting the cylinder closer to the 'scape wheel. In these very small and cheap watches side play of pivots causes great inequality of action.

PRACTICAL HINTS

Make it a rule to test every watch you handle for magnetism. A small pocket compass placed close to the balance when the watch is running will indicate by a vibrating motion if the balance is polarised, and, if it is, the watch and case should be treated in a de-magnetiser, to remove this trouble.

When repairing a watch inspect the balance pivots carefully to see that they are straight and in good condition. Examine the endstones, and if they show any wear it should be polished off, by using a small lap made of tortoise-shell about 30 mm. in diameter, mounted in the lathe, and a small amount of fine diamond powder mixed with oil put on the face of it. By holding the pitted endstone against this with a slight pressure, while the lap is running at a fairly high speed, it can be made as good as a new one in a very short time.

After this operation it is important to clean the endstone and setting thoroughly.

Examine the balance to see that it is true and in poise.

Do not open the bankings carelessly. Remember that the result of excessive slide is a dead loss of power, and this loss increases rapidly with any deterioration of the oil on the pallet stones.

Do not neglect to try the jewel pin to see if it is set firmly. Even a slightly loose jewel pin is a fruitful source of trouble.

Do not open the curb pins on the regulator. The hairspring should fit between the pins, without pinching, and without play, to get the best result in timing.

See that the hairspring is centred and flat, and has a sufficient amount of clearance under all conditions. Bear in mind that its regular vibrations will be increased a good deal at times, when the watch is subjected to sudden motions or shocks.

Do not neglect to remove any finger-mark or greasy matter from the plates, caused by the handling of the movement. For this purpose I find a buff stick very useful—a flat stick of wood, about 14 mm. wide, covered on one side with buckskin, such as is used for buffing. The end of this is dipped in benzine, wiped off rapidly with a clean cloth, and used immediately for cleaning off the top surface of the plates.

Do not expect a position adjusted watch to rate the same as it did originally after any change or alteration has been made in the balance pivots, or balance jewels. Even when the work is done with the greatest care this kind of repair may call for readjusting the movement, and this should be done by a watchmaker experienced in this class of work.

Do not consider it a bad investment to put as much money as you can afford into up-to-date tools. And do not consider the time wasted which you spend in keeping your tools in good condition.

Do not neglect to keep abreast of the times by reading good books and papers pertaining to the trade.

I wish to emphasise to the young watchmaker the importance of practice or training in the various branches of his work; and would recommend, as a profitable way of spending some of his leisure time, to take, for example, a discarded balance, and bend it out of shape, and true and poise it repeatedly for the purpose of gaining experience. We might state that although a beginner may work on a balance all day, and still not succeed in getting it in very good order, an expert can do 20 to 25 in an hour, and get them all good. This applies equally well to the work on the hairspring, the escapement, the pivots, jewelling, and so on. And I would also state that nothing but hard work and conscientious application to the work, coupled with a certain amount of study, will ever bring forth a skilful and efficient workman.

How to Set a Jewel in a Watch Plate.—To replace a broken jewel which is set directly in the plate, first carefully remove all the pieces of the old jewel, mount the plate with the hole true in a pump centre jewelling head, or a universal head. Run the lathe slowly, and with a pointed burnisher carefully open the old bezel until the sides are parallel, and the diameter about the same as the original size. If the bezel is made any larger than the original size it is very likely to break. The usual way to put the new jewel in place is to wet the end of the finger and touch it to the jewel, which will adhere readily. Push it into the setting and slide the finger off. This will leave the bezel and jewel wet, and hold the jewel while the bezel is being closed over it.

The form of tool used for burnishing down the bezel is snown in Fig. 136. This tool is held firmly on the T-rest at the proper height, the bevelled side towards the hole, as shown in Fig. 136. With the lathe running slowly, it is forced towards the plate until the bezel is closed firmly over the edge of the jewel as in Fig. 136.

If the bezel is broken, a jewel of the same outside diameter as the original one may sometimes be set securely by making a new bezel of larger diameter and burnishing it in far enough to cover the edge of the jewel. This is done by starting the burnisher at a point farther way from the jewel, as shown in Fig. 136. If the metal around the hole is cut away to such an extent that this method is impracticable, it will be necessary to either select a jewel of larger outside diameter, or put a metal bushing in the plate large enough to make a new setting of the diameter required for the new jewel.

To Remove a Broken Screw from a Watch Plate.— After all other screws and steelwork are removed suspend the plate on a copper wire in a saturated solution of alum in water. If the screw which is to be removed is very long it is necessary to take the plate from the solution every 24 hours, and with a sharp point remove the dissolved steel, in order to hasten the action of the solution.

Waltham Screw Taps.—The table on page 160 gives the pitch, the diameter of the thread and also the proper size of the tap drill, for all the screws used in the Waltham Watch movements. Taps of all these sizes are furnished by the Company at a nominal price.

To find which is the right tap for any of the screws used in Waltham watch movements, measure the diameter of the screw and refer to the table. If the screw measures ·65 for example, we find that 65 in the column of diameters of threads corresponds to No. 17 tap, for which the tap drill should measure ·54.

The only threads which we are unable to identify by measuring their diameters are Nos. 7 and 13, but in this instance the difference in pitch is sufficient to show at a glance in comparing one with the other.

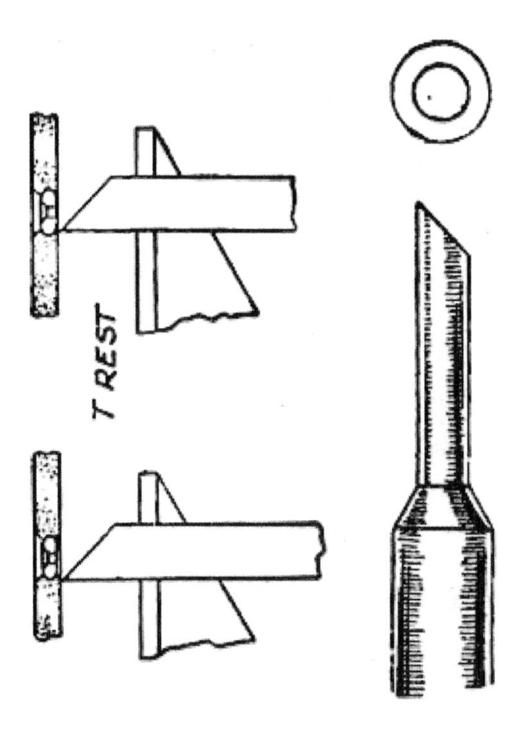

FIG. 136.—METHOD OF SETTING A JEWEL IN A WATCH PLATE.

Mainspring Don'ts.—DON'T fail to provide yourself with the best mainspring winder that can be obtained. See that the hooks on all arbors of the winders are no longer than the thickness of the thinnest spring, and thus avoid kinking, and, therefore, unnecessary breaking of mainsprings.

DON'T use a mainspring that is too long, because it fills the barrel and prevents that part, or the mainwheel, from making the required number of revolutions, with the consequence that the watch will not run as long as it should after each winding.

DON'T use a mainspring that is too strong, because it will set, increase the chances of breakage and injure the watch.

DON'T use a mainspring that is too wide, and be sure that the tip and brace do not extend beyond the limits of the cover and barrel.

DON'T forget that a mainspring should not occupy more than one-third the diameter of the barrel, thus leaving two-thirds to be divided between the arbor and winding space, to enable the watch to run about 36 hours.

DON'T expect a mainspring to be flat if you put it in the barrel with the fingers. This method usually injures the spring, gives it a conical form, and thereby increases the friction in the barrel.

DON'T bend the inner or outer end of the mainspring with flat-nosed pliers, but provide yourself with specially made round-nosed pliers which will give a circular form to these parts, prevent short bends, contract the inner coil, and thus secure a closer fit to the barrel arbor without injuring the spring.

DON'T expect other than a properly fitted flat mainspring with rounded edges to produce the least friction in the barrel, allow the greatest amount of power to the train, and give the best results as to time, service, etc.

DON'T expect a mainspring to always endure extreme changes in temperature, or electrical disturbances, or straightening at full length, or neglect from lack of cleaning and oiling.

DON'T expect a watch that needs cleaning or other repairs to run satisfactorily by merely putting in a new mainspring.

DON'T expect a mainspring to plough through too much dirt.

DISMANTLING: FITTING WINDING SHAFTS

As soon as a watch comes in for repair it should be thoroughly examined. It is a good plan to ask the owner a few questions about the past history of the watch. Answers to such questions as "Has it kept time in all positions?" "Have you had a new mainspring fitted recently?" will help towards a quicker diagnosis.

Before opening the bezel, examine the glass, which should be a perfectly tight fit. Frequently the bezel is fitted with an unsuitable glass. If the bezel is one with only a narrow inner flange to keep it away from the dial it will need a higher glass than one with a deep inner flange. Fig. 84 shows various kinds of watch glass.

Glasses.—It is surprising how many glasses carry a circular scratch made by the minute hand. Usually this indicates a low glass. By placing the nail upon the glass above the minute hand some idea can be gathered of the distance between the hand and the glass. If the watch is not working the ear test is quite effective. Pull out the winder to the set hands position place the glass close to the ear and turn the winder. A light scraping noise will be heard if the hand is touching the glass. With small watches this test may not be effective, so something more positive is necessary. Mix a spot of rouge with a spot of oil and apply it to the highest point of the minute hand, snap down the bezel, and set the hands. Open the bezel and examine the glass. If there is a red ring it will prove that the hand is too high. The hands should be both parallel to each other and to the dial in all positions.

Thin Cases.—Modern cases are tragically thin. They should be carefully inspected for broken hinges, cracks, torn loops and last but not least tiny pin-holes in the bezel. Mirage and other fancy shaped bezels having a number of sharp corners give endless trouble in this way. The holes are often very minute, but they allow a considerable amount of dust to enter, and unless they are covered it is useless to clean the movement. The easiest way to fill the holes is to remove the glass, apply a little flux and a tiny piece of soft or tinman's solder at each hole and gently warm the bezel over a low gas or spirit flame. Well oil the hinge before warming the bezel for the heat may affect its action.

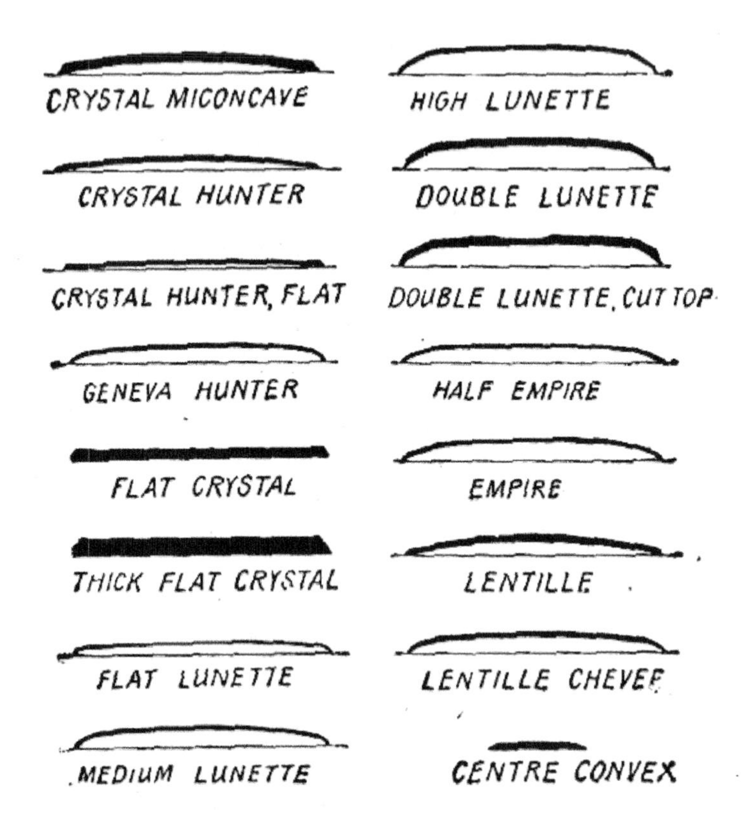

FIG. 84.—VARIOUS TYPES OF WATCH-GLASSES.

The winding button and shaft should also receive attention. If this is worn, there will be plenty of room for dirt and dust to enter through the shaft hole. As well as the possibility of losing the complete winder undue strain will be placed upon the shaft and the internal winding operation and hand setting mechanism. Continuous operation of the winder in this condition will finally end in a breakage. If the wear is excessive a new oversize shaft will be necessary. First remove the dial by unscrewing the dial feet screws. In some watches the screws are placed in the edge of the bottom plate and only accessible when the movement has been removed from the case. In others, the screws are provided with a semicircular flange at the base and screwed into the top surface of the bottom plate. If the minute hand is a good fit take care in levering it off, otherwise it may jump away. If the "seconds" hand is a tight fit, carefully raise the dial which will also act as a lever and remove the hand. The hour hand will come away with the dial, bringing with it the hour wheel.

Dismantling the Works.—When the dial has been removed unscrew the cover piece which keeps the spring return lever—a small lever working in the groove of the castle wheel—and the hand-setting wheel in place, remove the castle and crown wheels, and everything is ready for fitting the new winding shaft. Fig. 85 shows the crown wheel, and castle wheel. Select a "rough"—in other words a partly finished winding shaft—that is a little larger than the hole between the plates, as most of the wear will have taken place here and in the pivot hole. With the old shaft as a pattern mark off the length to the end of the pivot. Place the shaft in a suitable lathe chuck and turn down the shoulder and the pivot. When the new shaft has been turned down a little try the fit. To do this there is no need to remove the shaft from the chuck, simply take the plates to the shaft and there will be every opportunity of making a moderately tight fit. Leave the shaft a little tight to turn as final buffing will reduce it to a smooth action.

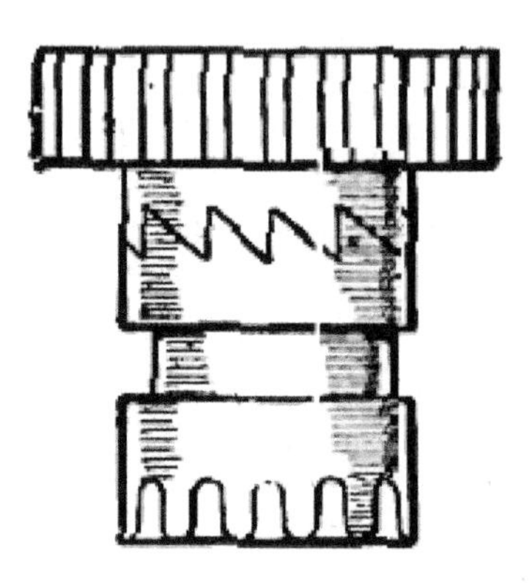

FIG. 85.—THE CROWN AND CASTLE WHEELS.

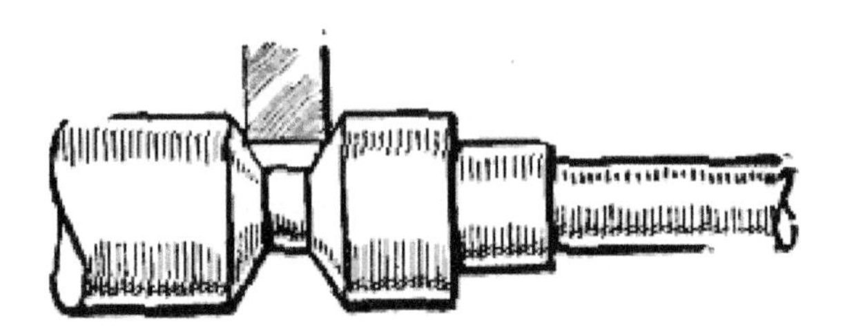

FIG. 86.—THE EFFECT OF A BADLY MADE BOLT GROOVE.

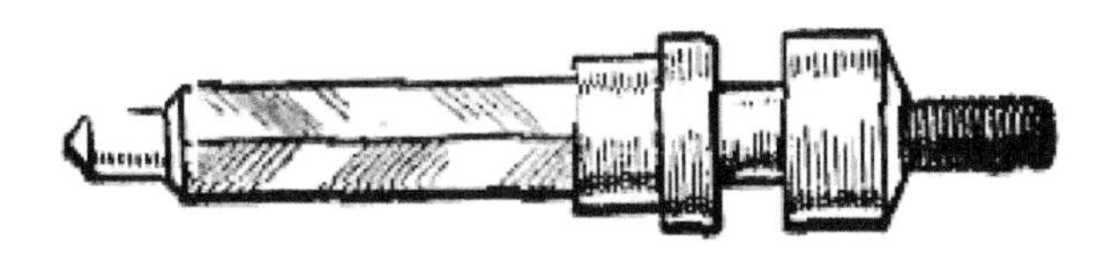

FIG. 87.—A FINISHED SHAFT.

The Crown Wheel.—As the squared part of the shaft is always smaller than the shoulder for the crown wheel it will be quite safe to mark the position of the crown wheel shoulder and turn the shaft down until it just fits the crown wheel. When the shaft has been turned to fit the crown wheel, remove it from the lathe chuck and place it in a hand chuck. It is most important that the squared section should be square. Failure to do this may cause the castle wheel to have an uncertain action. Most watchmakers use a boxwood block for filing purposes—a small cube of box wood. Although box is a hard wood, steel pieces will become firmly embedded.

If you feel a little uncertain with regard to your filing capabilities, reverse the crown wheel and place it upon the shaft. It will act as a guide and prevent filing the crown wheel shoulder. Use a fine file with a safety edge. Try the castle wheel on each square to make sure the action is smooth. When the square is finished replace the winding wheels in the watch, and put the shaft in place and try the forward and backward action. Lightly screw up the pull-up or bolt piece which retains the shaft, and mark the position of the slot. Care should be taken in making the sides of the slot straight as a V-shaped slot will be inclined to force the bolt out. Fig. 86 shows the effect of a badly made slot.

Sharp Corners.—Whilst the shaft is still in the lathe turn the pivot to a point as this will make entry easier. The corners of the lower end of the square can be turned off too. Sharp corners have a nasty habit of cutting away the brass plate and upsetting the winding. Put the winding shaft in place and screw the bolt piece tight. Mark off the position of the end of the shaft a little beyond the pendant—the tubular extension on the side of the case—remove the shaft and cut it off. Always remove a shaft before cutting it off, as the shock is quite sufficient to break a balance pivot. If the end is already screwed, attach the winding button. If the end is plain turn it down to a suitable size for threading. Fig. 87 shows a finished shaft.

Interchangeable shafts are supplied by the material dealers for most of the modern watches. If ordering by post, the size of the movement and the make, or preferably the broken part of the old shaft if available, should be sent to the dealer.

FITTING MAINSPRINGS

Although the mainspring is a very essential part of a watch, it is frequently treated by repairers as though it were of a minor rather than a major importance. The life of a mainspring is by no means everlasting. In fact, many mainsprings have a decidedly short life. This makes it all the more difficult to convince a watch owner that a new mainspring is necessary when the watch is working with the existing mainspring.

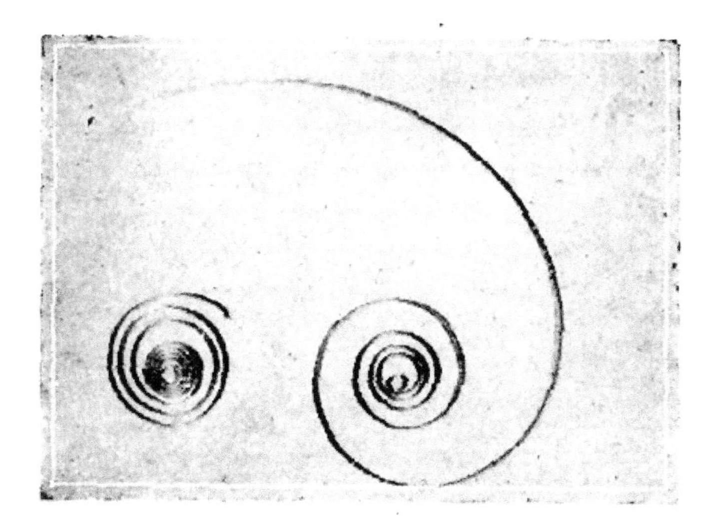

FIG. 88.—SHOWING AN OLD AND NEW SPRING. THAT ON THE LEFT HAS BEEN IN USE FOR A CONSIDERABLE TIME.

Removing a Spring.—When a mainspring is run down there should be as much unoccupied space as the spring occupies when it is lying against the side of the barrel. The space in the barrel should be allotted equally to the mainspring, the barrel arbor and the unoccupied space—one-third each. When removing a spring from a barrel, pull up the inner coil carefully and allow the spring slowly to unwind itself from the barrel against your fingers. To let the spring literally fly out from the barrel will be courting disaster, as the necessary entanglement of the coils will distort the spring and render it useless. Loss of power is often difficult to discover when it is not constant. It is very important therefore to make a careful examination of the spring when it has been removed.

The appearance of bright spots will indicate friction between the coils with a consequent jerky or unequal pull. Oil will have little effect upon coil friction, and the balance action will be certain to "fall off" at the weak pull. The only remedy will be a new mainspring. If there are bright marks on the edge of the spring, and the barrel cover shows a series of circular scratches, this will indicate that the spring is fouling the cover, a frequent cause of loss of power. One remedy is to reduce the thickness of the barrel cover by placing it in a lathe step chuck and turning off some of the excess metal, taking care not to remove any of the arbor bearing in the operation. If the cover is too thin to reduce, a new spring, a size lower, will have to be fitted.

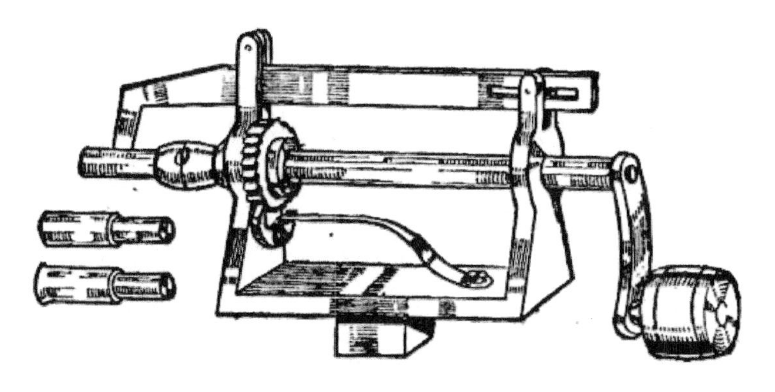

FIG. 89.—A POPULAR MAINSPRING WINDER.

Faulty Mainspring.—The fact that a watch goes does not signify that it is going correctly. The escapement and all the other parts may be in perfect adjustment, but still the balance may not have the action it should. In this case the fault can usually be traced to the mainspring. The coils should not be disturbed any more than is absolutely necessary, which probably accounts for this attitude of indifference towards the mainspring. If the coils are very gummy the spring should be soaked in benzine, and then allowed to dry off. Any slight stickiness can be removed by passing a piece of folded tissue paper between the coils with the aid of a pair of tweezers.

Fig. 88 is an actual photograph of two mainsprings (an old spring and a new spring). That on the left has been in use for a considerable time, and has become "dead" in the centre. In other words the inner coils have lost a good deal of their elasticity. If this spring were fully wound it would probably give the balance a fair action for the first 12 hours; but after that there would be a considerable "falling off" in the action, which would seriously affect the timekeeping. Such a spring is really unfit for further use.

The Barrel Arbor.—The barrel arbor should be a good fit in both barrel and cover, with only the minimum of endshake. As the mainspring exerts a certain amount of twisting force when wound, an ill-fitting barrel will also be inclined to twist and undoubtedly come into contact with the underside of the centre wheel with disastrous results. If either of the holes need re-bushing, broach or reamer the hole round, turn a small stopping of brass or nickel and rivet it in position. Place the barrel or cover in a step chuck, turn off any surplus rivet, centre the new bush and drill a true hole. Finally broach the hole to fit the arbor and chamfer a small oil sink.

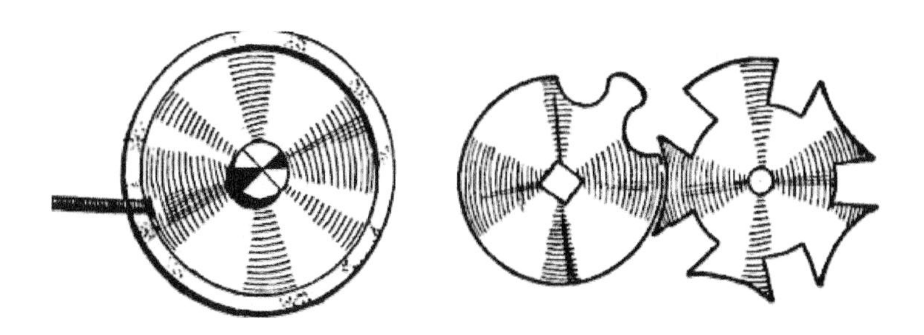

FIGS. 90 & 91.—A NEW HOOK BEFORE BEING CUT OFF. (RIGHT).—THE STOP WORK MECHANISM TO PREVENT OVERWINDING, AND ALSO TO ENSURE THAT THE SPRING DOES NOT RUN DOWN BEYOND A CERTAIN POINT, THUS MAKING FOR BETTER TIMEKEEPING, AND EVEN TENSION THROUGHOUT THE 24 HOURS.

The safety of the mainspring depends upon a good hooking attachment. If the barrel hook is one of the screwed or riveted type, it should be perfectly rigid. It should never stand out from one side of the barrel more than the thickness of the spring. The smaller the hook the better; a large hook will not only be more difficult to fix, but it will occupy more space in the barrel. To fit a new hook, drill and tap a small hole through the side of the barrel at an angle. This will give a longer hole and allow more threads.

File a piece of steel wire and tap it until it shows a full thread, cut off a little above the full thread, and file the top and two sides flat. Screw the new hock from the inside to the outside until only the head stands out from the side of the barrel, cut off the surplus and file flat (Fig. 90). It will be an advantage to undercut the hooking side of the head with a slitting file.

The Barrel Hook.—Modern watches favour the recessed type of barrel hook—a step cut in the side of the barrel—as there is no risk of undue projection. As this kind of hook is very shallow it must be square cut to prevent the mainspring from slipping. Any sign of a rounded nose can be rectified by using a sharp, long-pointed graver. Many barrels have a hook which has been pressed through the side with a special tool, and many mainspring punches have a cutter for this purpose, but the risk of distorting the barrel with one of these punches is so great that the screwed-in hook is a much better proposition.

Types of Spring.—When replacing a worn or broken mainspring it should not be assumed that the existing spring is the original or even the correct kind of spring for the watch. It may be too weak or too strong. With a strong spring the balance would have an excessive vibration when fully wound with a sudden decline after a few hours, whilst at the end of 24 hours running the balance would probably come to rest if the watch was placed in a "pendant" up position. Such a condition is most undesirable. A weaker spring would give a smaller but more constant vibration and serve the same purpose as a stop work. The purpose of a stop work, which consists of a finger-piece attached to the barrel arbor engaging with a star wheel screwed to the barrel or barrel covers, is to allow only the middle turns of the mainspring to be used.

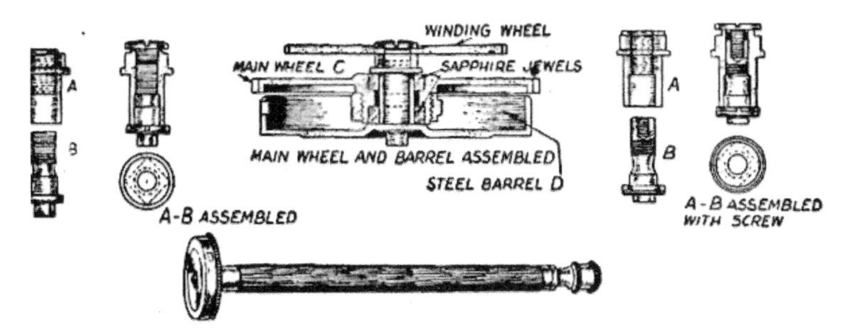

FIG. 92.—THE WALTHAM MAINSPRING BARREL. IT IS OF THE "FREE" TYPE. BELOW, THE SPECIAL TOOL FOR SEPARATING THE BARREL.

The stop-work allows only four turns, so that a spring which makes six turns will permit the stop-work being set up one complete turn, and still allow one turn unused. Fig. 91 shows a stop-work mechanism. From this it will be gathered that at least four turns of spring are necessary. At least five turns should be aimed at whether or not the watch is fitted with a stop-work, as this will give a little reserve. When the length of spring has been determined the outside end will have to be fitted with a hook according to the type of barrel. If a stop-work is fitted, an ordinary hole will be quite suitable, as the spring will never be pulled completely away from the side of the barrel. Before punching or drilling the hole remove the excess hardness from the end of the spring by "letting it down" over a spirit flame. Run a broach through at an angle to make a sharp hooking edge, file the end round and finish with an emery buff. Watches not fitted with a stop-work need a more resilient form of hooking as the mainspring is pulled well away from the side of the barrel when fully wound.

Pivoted Brace Hook.—Many pocket watches use the pivoted brace hook. This consists of a small tongue with two pivots, riveted to the end of the spring. The pivots fit in holes drilled in the cover and bottom of the barrel. The pivoted brace allows the spring to be tightly wound without undue strain on the extreme end. When fitting a pivoted brace hook make sure that it does not exceed the width of the spring, or what is probably worse, be out of alignment with the spring. Fig. 93 shows two kinds of plain hook and the pivoted brace.

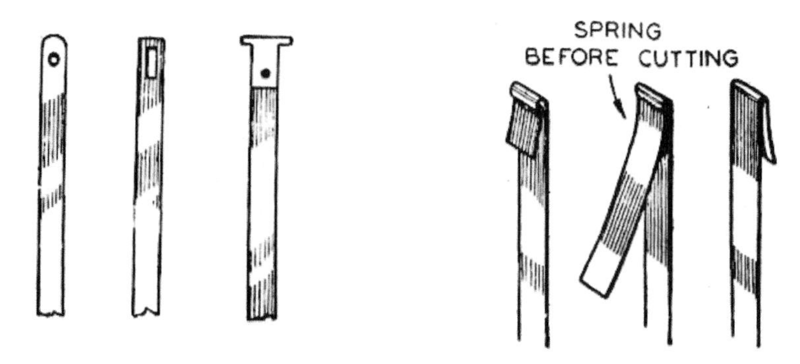

FIGS. 93 & 94.—(LEFT)—TWO KINDS, OF PLAIN HOOK AND THE PIVOTED BRACE; (RIGHT) TYPES

OF BENT HOOK.

FIG. 95.—HOW THE SPRING END GRIPS THE RECESS IN THE MAINSPRING BARREL.

Plain Bent Overhook.—Quite 50 per cent of the mainsprings fitted by the manufacturers have a plain bent overhook, made by heating the end of the spring, doubling it back and squeezing it flat with a pair of pliers. Although the manufacturers use this kind of hook it is not one of the best. There is too much strain at the actual bend. Ample proof is to be found in the number of springs which break at this particular point. A much more satisfactory hook is one in which a loose piece of spring is inserted between the bent end of the mainspring and the barrel. To make a hook, gently warm the spring at the point at which it is to be bent (the waste part will bend itself back). As the spring becomes red, squeeze it gently with a pair of pliers. Break off a small piece of waste spring, place it between the bend, warm the end again until it becomes red and squeeze tightly. The squeezing must be done whilst the spring is red, or there will be a tendency to break.

When cold, cut off the waste spring, using the edge of a triangular or square file to make the cut, leaving just a small bend. File off the corners and buff off the discoloured part of the spring. Cut off another small piece of waste spring (there are usually 2 or 3 inches to spare) and insert it in the hook. Fig. 94 shows types of bent hook. Fingers should never be used to wind a spring into a barrel. Apart from fingermarking, which is likely to lead to rust, the spring will assume a spiral shape causing friction with the barrel cover and a consequent loss of power. Wind the spring on the mainspring winder arbor first, place the barrel over the spring, reverse the ratchet on the winder and let the spring unwind itself into the barrel. Fig. 89 shows a popular mainspring winder.

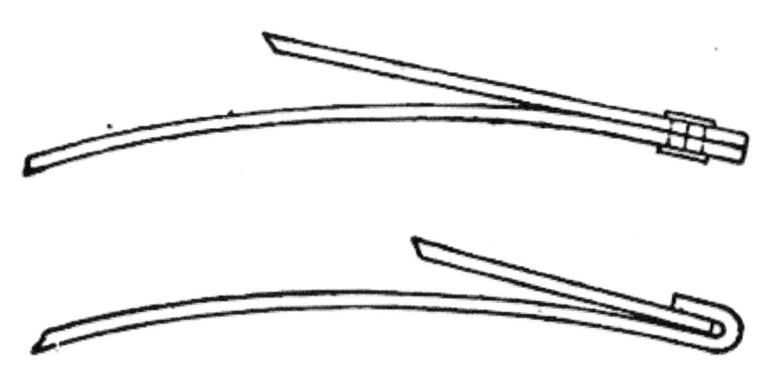

FIG. 95A.—TWO TYPES OF MAINSPRING END—

THE RIVELED AND THE LOOSE END.

Special springs and springs with patent hooking devices should be specially obtained. In fact, the spring designed by the makers should be used whenever possible, as it will provide the maximum of power.

BALANCES AND HAIRSPRINGS, ADJUSTING AND TIMING.

A Balance is a fly-wheel, and may be circular or any other shape; it may run true or not. The one essential is that it must be in perfect poise; that is, it must have no heavy part, and when rested on the straight edges of a poising tool it must have no tendency to settle in any one position. It is an advantage to have as much of the weight of a balance as possible in its outer rim. It is also an advantage, principally for the sake of appearance and for convenience, that a balance should be circular and should run fairly true.

Balances may be plain or compensated. Plain balances are made of gold (so as to be non-corrosive and keep clean), brass, or steel. They generally have three light arms and a true circular rim.

Temperature Error.—A balance is controlled by its attached hairspring, and the time of its vibrations depends upon the strength of the hairspring. A strong spring will cause rapid vibration, and a weak spring a slow motion. Both hairspring and balance are affected by changes in the temperature. The balance expands as it is warmed, and contracts as it is cooled. When it expands the arms lengthen and the rim increases in size, removing the weight further from its centre and causing it to move more slowly. When a spring is warmed it loses some of its force, also causing the balance to move more slowly. The combined effect of a change of temperature of 45° Fahr. (from the cold of a bedroom dressing-table to a warm waistcoat pocket, say from 40° to 85°) is to cause an uncompensated watch to lose several minutes per day. Temperature also affects the depths a little by causing the wheels to expand; it alters the strength of the mainspring, and affects the fluidity of the oil. The combined effect of all these is the "temperature error" of the particular watch in question, and varies in each individual watch. Therefore machine-made watches cannot be turned out "compensated," but each one must be finally adjusted by itself.

Compensation Balance.—The net result of a change of temperature is to cause a serious loss in heat and a gain in cold. To counteract this the compensation balance is used. Fig. 147 shows the construction of an ordinary compensation balance. It is a circular rim with a steel crossbar. The rim is bi-metallic, being steel inside and brass outside, and cut through at two opposite points. Brass is more affected by heat than steel, and in a rise of the temperature the outer brass will lengthen more than the inside steel. The effect of this is to curve each half of the rim inwards and bring the weight of the balance as a whole nearer to its centre. This causes the watch to go faster, and, if the amount of the inward movement of the rim is exactly sufficient, will compensate the tendency of the watch to lose. These balances are weighted by screws fitted in a series of tapped holes all round the rim. By moving the screws nearer to the free ends of the two segments of the rim the effect of the balance is increased; by moving them towards the fixed portions the effect is diminished. Therefore adjusting for temperature consists in trying the watch in cold, then in heat, and moving the screws according to the performance of the watch, until its rate in cold (40° to 50°) is equal to its rate in heat (80° to 90°).

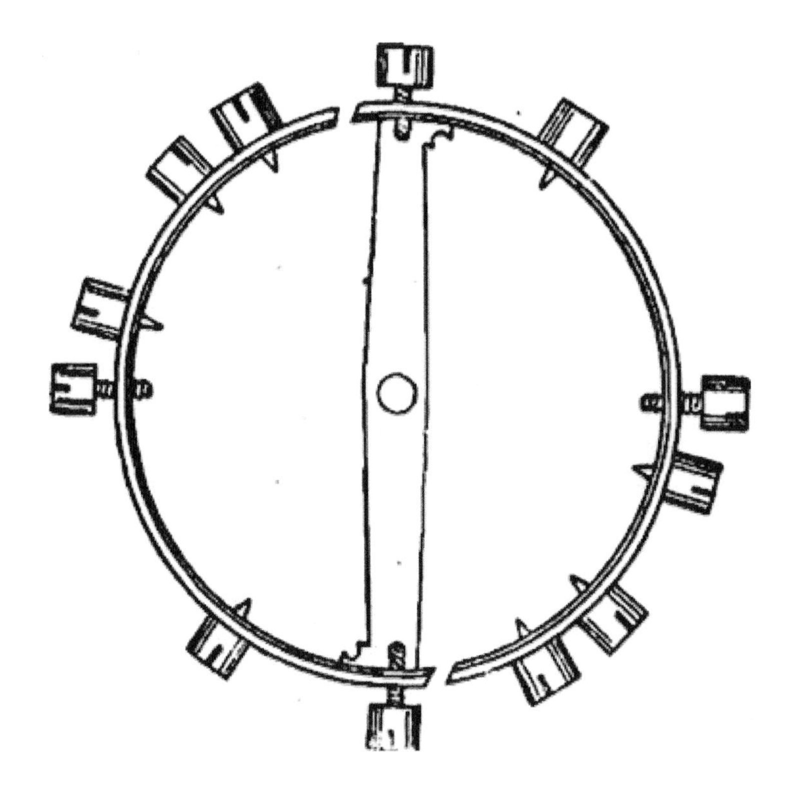

FIG. 147.—COMPENSATION BALANCE.

In Fig. 147 there will be noticed four screws at equal distances from each other, with long taps. These are the "quarter screws." They are never moved for temperature adjustment, but are for poising the balance or for small timing alterations. Drawing one out a little makes that part of the rim heavier. Drawing out an opposite pair will slow the watch. Turning a pair in will make it go faster. In the best balances "quarter nuts" are fitted instead of screws. Fig. 148 shows a quarter nut. It is a gold nut, turning on a fixed steel screw, and is not so liable to work loose as a plain quarter screw from frequent turnings. A quarter nut is split, and slightly sprung on to its screw to move firmly and not get loose. To turn these nuts a split screwdriver blade like A (Fig. 148) is used.

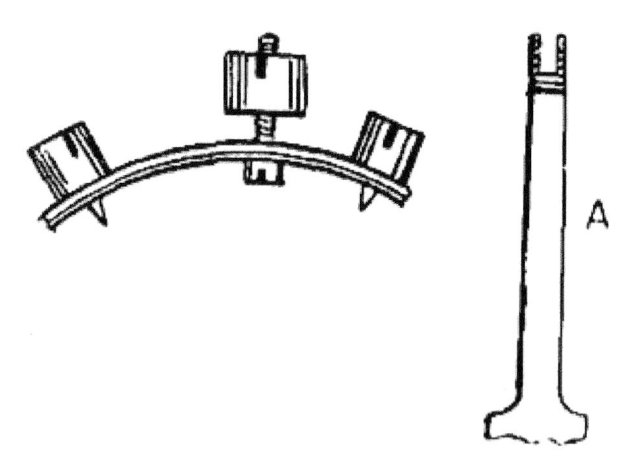

FIG. 148.—QUARTER NUT AND SCREWDRIVER.

In making a compensation balance, a steel disc is turned up true, with a central hole. It is then covered with molten brass, which adheres to it all over. The brass is filed off the flat sides of the disc and the central hole cleared; then the surplus brass is turned off the edge, leaving the thickness required. It is hammer-hardened and turned smooth. The interior of the balance is cut out by turning and filing, leaving the crossbar; the rim is drilled and tapped in a dividing engine, and finally cut through at two opposite points, after being mounted upon its staff by the escapement maker.

When a balance is cut the unequal hardness of the brass and steel composing its rim generally causes the two segments to go out of truth. They are trued by removing all the screws and putting in the turns to note where they depart from the circle, and bending with the fingers as far as possible or with brass pliers. Absolute truth looks very nice, but is not essential. Compensation balances that have been running some time generally go a little out, and, so long as they can be poised by means of the quarter screws, are best left alone. The only way to true them is to remove all the screws. Before doing this it is best to make a sketch of the balance rim, noting the position of the screws, so as to replace them as before.

A watch with an uncut compensation balance is no better than one with a plain balance; but if the balance be cut and trued as described it will be greatly improved. A watch with a cut compensation balance, not specially adjusted, like the great majority of ordinary watches, is cured of *most* of its temperature error, and may generally be depended upon not to vary more than 30 secs. from its rate in one day between 40° and 85°. An adjusted watch is one with a compensation balance cut, and the screws arranged by trial to reduce the error to about 2 sees, per day or less.

Hairsprings.—Fig. 149 shows an ordinary flat hairspring. Its outer end is pinned into a fixed stud and its inner end into a collet.

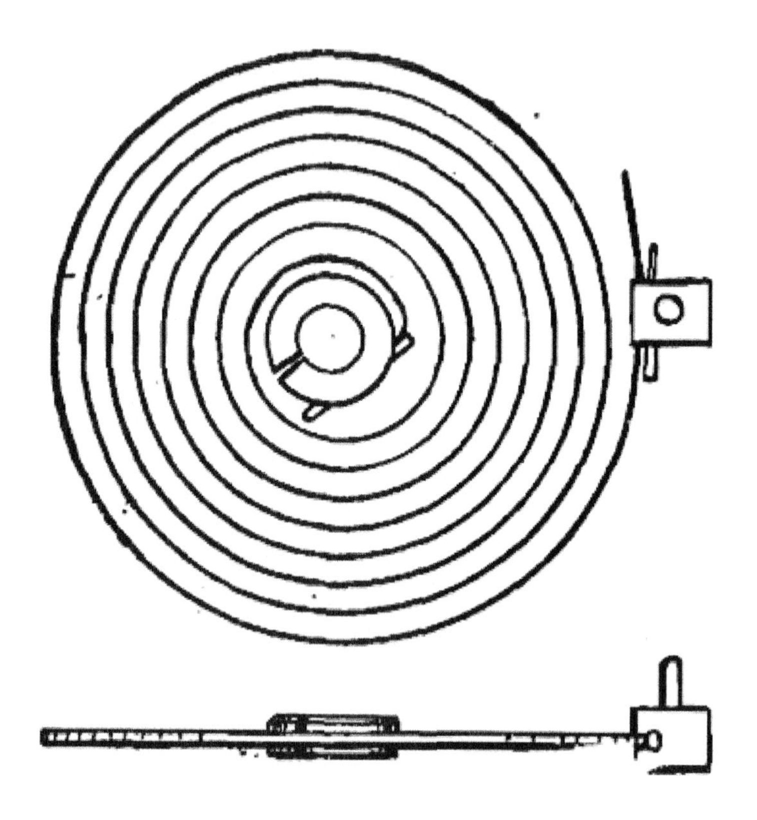

FIG. 149.—FLAT HAIRSPRING.

The stud may be a small square of brass fixed in the watch plate, as in some English full plates, a small square of brass pushed friction tight into the balance cock, as in many Genevas, or a steel stud screwed to the plate or the balance cock, as in the best English levers, American and Swiss watches. The collet is in most watches a small brass circle turned to fit the balance staff friction tight, and split to give it a certain amount of spring, as in Fig. 149. A better form of collet, used in the best English levers, is that shown in Fig.

150. It is a circle of steel, hardened and tempered, made to accurately fit the staff. It has two flats filed upon it, making it nearly oblong. Such a collet goes on more truly, and is not liable to become damaged; it also allows the hairspring to be trued in the centre more easily.

The spring itself is of steel wire, ribbon-shaped, drawn hard, and coiled up into a close spiral. The closeness of the coils depends upon how many are coiled up together. Thus, if three lengths of wire are coiled up together, when released the resulting spirals will be rather open. If two are treated together the coils will be closer. If only one is coiled up on itself the coils will be very close and nearly touch. The closer a spring is coiled the longer it is for a given diameter.

The strength of a spring depends on its thickness and width—that is, its stiffness—and its length. Hairsprings are sold in small packets, numbered according to their strength in a series of numbers which differ with each maker, and are only useful to compare one spring with another of the same make. It is, therefore, best in hairsprings to keep to one make only for ordinary quality springs, and to send specially for one of extra good quality when required.

To select a new hairspring, the number of beats required in an hour must first be known. This is termed the "train" of the watch. Most watches have 18,000, 16,200, or 14,400 trains; that is, they make that number of beats in one hour. All Genevas and American watches have 18,000 trains. Old

English watches often have 14,400 trains, and many more recent ones 16,200, or 15,400. But it is probable that all watches in the future will be made 18,000.

The train of a watch can be ascertained by multiplying the numbers of teeth in the centre, third, fourth, and scape wheels, and dividing the result by the third, fourth, and scape pinions. Twice this number is the train.

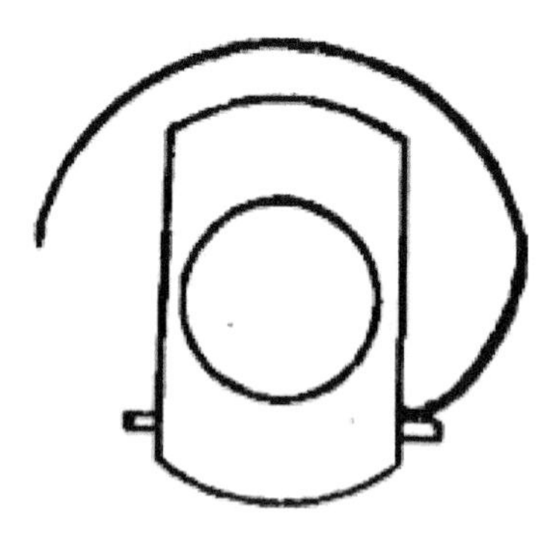

FIG. 150.—SQUARE STEEL COLLET.

The following table gives the usual trains found, with the number of beats in one hour, in one minute, and in 20 seconds.

Description of Watch.	Centre wheel.	Third wheel.	Fourth wheel.	Scape wheel.	Third pinion.	Fourth pinion.	Scape pinion	Train.	Beats per min.	Beats in 20 secs.
Geneva or American	80	75	70	15	10	10	7	18,000	300	100
"	80	75	80	15	10	10	8	18,000	300	100
"	64	60	60	15	8	8	6	18,000	300	100
English	80	75	80	15	10	10	8	18,000	300	100
"	64	60	70	15	8	8	7	18,000	300	100
"	64	60	63	15	8	8	7	16,200	270	90
"	64	60	60	15	8	8	7	15,400	257	86
"	64	60	56	15	8	8	7	14,400	240	80

It will be noticed that in all these trains the fourth wheel makes one revolution per minute; and when the fourth wheel has ten times as many teeth as the scape pinion, the train is 18,000; when nine times, the train is 16,200; and when eight times, the train is 14,400.

In counting vibrations when fitting hairsprings, double vibrations only are counted, thus each time the balance comes to the left is counted one. In this way an 18,000 train counts 75 in half a minute; a 16,200, 67; and a 14,400, 60; or in proportion for twenty seconds or a full minute.

FIG. 151.—COUNTING THE VIBRATIONS OF A BALANCE.

Pick out a spring that is a little too large in diameter to lie in the curb pins of the index. Lay the spring on the balance and press the collet down upon it to temporarily hold it in place. Hold the end in tweezers and let the balance hang down with its lower pivot resting on a watch glass. With a turn of the fingers and tweezers, set the balance vibrating about half a turn each way, or more. Then hold it perfectly still and steady and it will continue to vibrate for a full minute. The vibrations can be counted for twenty seconds to see if it is anywhere near what is required. Fig. 151 shows how the spring and balance are held. If the balance moves too slowly, select a stronger spring, if too fast, a weaker one, and try again. The spring should be held in the tweezers at the exact point at which it will have to be pinned in its stud; thus, if the spring is two coils too large to go in the curb pins, it must be counted while held two coils from the end.

Finally, select one that counts one double beat slow in a full minute. Lay this on a convex watch glass on white paper, and with tweezers and a needle point, or two pairs of tweezers, break out short pieces (about 1/4 turn) at a time from the "eye" or inner coil, gradually enlarging the central opening until it will go easily over the collet with a little room to spare. Then bend a short piece sharply inwards to pin in the collet, as in Fig. 152. For breaking out and bending springs, a needle set in a handle, filed up, and slotted as shown at A (Fig. 152) is useful.

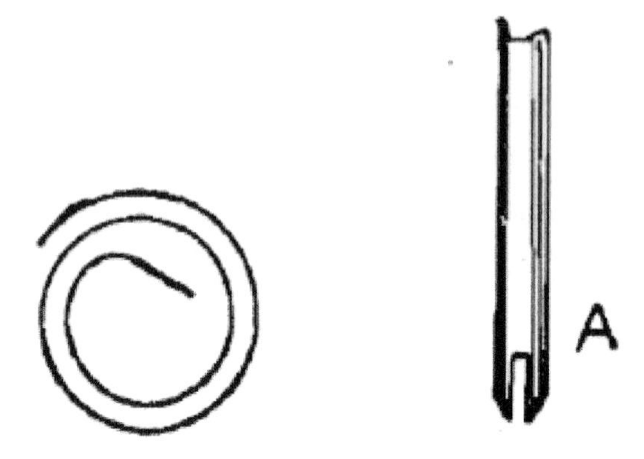

FIG. 152.—EYE OF HAIRSPRING.

To pin the spring in its collet, put the collet on a broach, pass the spring over the broach, and insert the end with tweezers. File up a fine burnished brass pin, and file a flat upon it, making it D-shaped. Insert this pin to see if it fits, with the flat against the spring. Cut off any that projects, and try again. Cut off the end until the pin goes in half-way through the hole. Then lay the pin on the filing block (still in the pin vice), and half cut it through with a knife; insert it finally, and break it off in the hole. With very strong tweezers press the pin well home. A hairspring must be pinned tight in its collet, or nothing can be done with it. When pinned in, put the collet and spring on a turning arbor in the turns and revolve with a light bow. Set it flat with tweezers so that it runs true. Note if it is true and concentric in the "eye," and in which direction it is out. Take it out and lay on the watch glass to bend true as required, try in the turns again, and so on until it is both flat and true. Put it on the balance and count for a full minute. If correct, break off the outer waste coils and pin it in the stud with a flatted pin. To set the spring central and flat on the balance cock or plate, pin it in as in Fig. 153, and bend it until the outer coil lies between the curb pins without strain, the spring stands level, and the collet is central with the jewel hole. If all these points are attended to before the spring is put in the watch at all, there will be no bending or cramping needed afterwards, and the balance and spring will go straight in, lie true and flat, be central, and free of everything.

The watch can be started and set by the seconds hand with the regulator clock. A loss or gain will be quickly seen. If too slow, take the spring up a little and re-pin in the stud; if too fast, let it out. It is here that the advantage of counting the spring one beat slow is found. If the balance is a compensation, or a sham one with screws, to make it go faster, a pair of screws can be reduced by filing or drawn out altogether. If too fast, a pair of screws can be added, or for a small alteration, "timing washers" can be added under the screw-heads. These are small washers stamped from thin brass sheet, and can be bought by the gross.

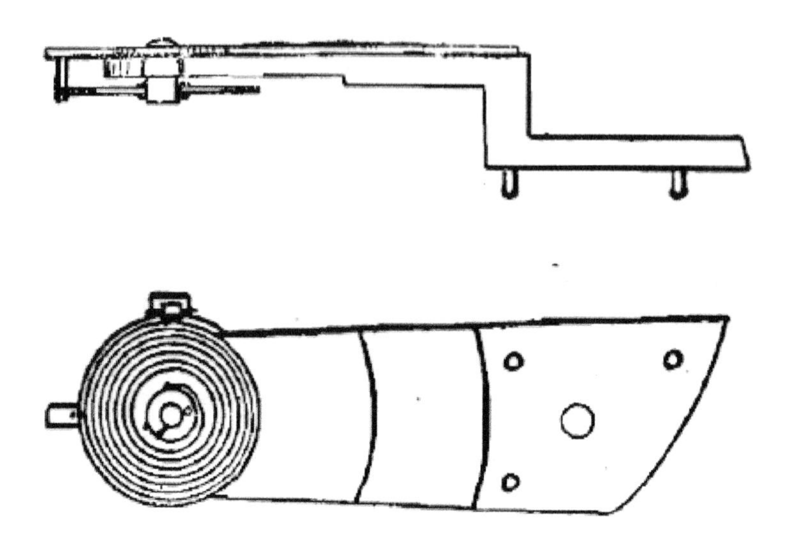

FIG. 153.—SETTING A SPRING TRUE AND FLAT.

Balance screws can be reduced by holding in a small pin vice of the pattern shown in Fig. 154, and revolving it in the fingers as the file is passed across the screw-head. Smooth them with a 3/0 emery buff and polish with a rouge-on-leather buff. Always be careful to leave the balance in perfect poise after any alteration.

FIG. 154.—SMALL PIN VICE.

The hairspring and its collet can be removed from the balance staff by levering up with a sharp pocket-knife alternately from either side. A collet can be turned round by inserting the thin blade of an oiler into its slit and using it as a lever.

A flat hairspring should be pinned in at equal turns, as shown in Fig. 149; that is, it should consist of so many complete turns, finishing against the point where it starts from the collet. Such springs time better in good watches.

Sometimes when a flat hairspring is pinned in its collet and in its stud it will not lie flat. It is dome-shaped or cup-shaped, owing to the stud hole not being drilled level with the collet hole. In such a case the innermost coil of the spring, just where it leaves the collet, must be bent up or down as required to get the spring flat, as shown at A (Fig. 155), and the spring then set flat and true in the turns.

FIG. 155.—SETTING A HAIRSPRING.

In some very flat Geneva watches and undersprung English levers there is very little room for a hairspring without it touching the plate, the balance arms, the stud, or the index. For these watches a spring made from specially *narrow* wire must be got from the hairspring makers.

To lower a collet and spring bodily, or to lower the top of a collet to prevent it fouling the balance cock, the collet, together with the spring, may be placed on an arbor and turned down. A very sharp graver must be used and light cuts.

Breguet Springs.—Fig. 153 shows a breguet or overcoil spring at A. This is perhaps the best form of hairspring for any watch. Sometimes a double overcoil is made, as at B, but apparently has no advantage. Helical or cylindrical springs, as at C, are used in marine chronometers and some pocket watches, but do not seem to perform any better than a single overcoil breguet.

Breguet springs are made from flat springs by the workman who springs the watch. For this purpose hardened and tempered hairsprings are best. These can be obtained by sending the balance to the spring maker and telling him the train. Sometimes springs of palladium wire are used, especially in "non-magnetic" watches. This metal does not rust, and cannot be magnetized. Balances also are made with palladium in place of steel for the same purpose.

Having procured or selected a spring, preferably a close coiled one, with a diameter equal to half that of the balance rim, proceed to put it on like a flat spring, getting it *perfectly* true in the eye and flat. Count it exactly to time, and break off all surplus coils.

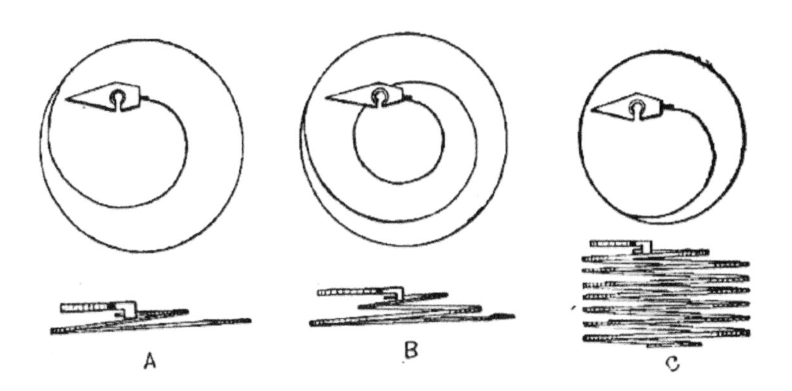

FIG. 156.—FORMS OF HAIRSPRINGS.

Unless a breguet spring is quite true in the eye it looks very bad and will not time well. When a spring is true there will be seen one coil, about midway between the outer coil and the eye, that apparently stands still. This is, of course, only an optical illusion. If a spiral be revolved in one direction, the coils all appear to run outwards; in the other direction they will all seem to go in. The "stationary coil" is merely that point at which this optical effect is exactly neutralized by the mechanical opening and closing of the spring caused by the vibration of the balance and which takes place in the opposite direction to the optical effect However, this coil should be seen in a true spring, and should apparently lie motionless. If it jumps or shakes, the eye of the spring is not quite true.

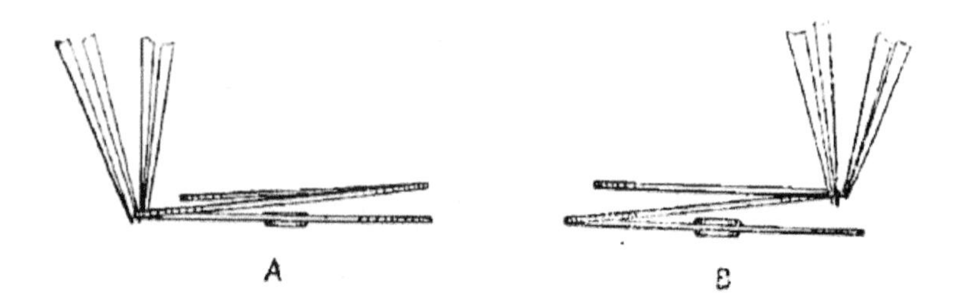

FIG. 157.—BENDING UP AN OVERCOIL.

When broken down to size, take two pairs of tweezers and bend up the outer coil like Fig. 157, A. Hold the spring firmly with one pair and twist the outer coil upwards with the other pair, in a gradually ascending slant. Halfway round the raised coil it will want another twist up, to make the raised part level, as at B. Take a curved pair of tweezers, like Fig. 158, and proceed to curve the raised coil inwards by nipping it up tight. Complete the overcoil with two pairs of ordinary tweezers, to the form shown in Fig. 156, at A. See that the overcoil is quite free from the second coil at the point where it commences to curve inwards. To judge the height required for the overcoil, put the balance in the watch and see how far up the balance staff the level of the pin hole in the stud comes, by sighting it across. Raise the overcoil to this point. If the watch has an index, the last quarter turn of the overcoil must be circular, and of the same radius as the curb pins.

FIG. 158.—CURVED TWEEZERS.

This can be measured with the millimetre gauge. All manipulation of the spring should be done while on a watch glass over white paper. Finally, pin the spring in its stud with a flatted pin, and drive it home hard. With a breguet spring, all timing is done by means of the balance screws, and the spring itself is not disturbed. Set it flat and true as it lies on the cock, so that it stands level and with the collet central over the jewel hole. Sight this across in two directions, as in Fig. 159. Then put it in the watch with the balance and see if it is flat and correct. If the spring is hollow the overcoil is not high enough, and must be raised a little; if domed it is too high.

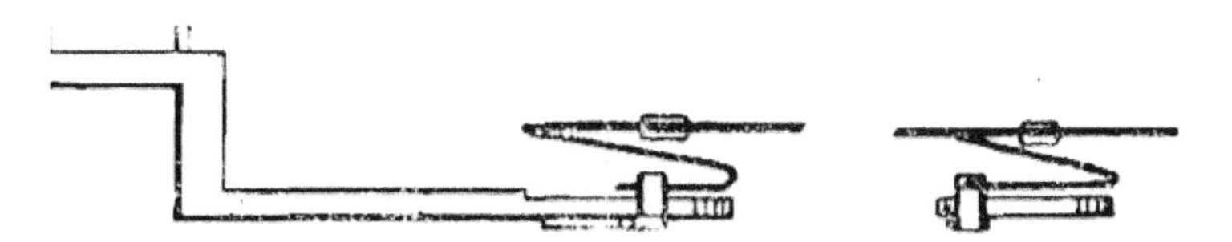

FIG. 159.—SETTING A BREGUET SPRING FLAT.

A great deal has been said and written about breguet springs being pinned in at equal turns, like Fig. 160, so that the spring starts at the eye at the same point or in the same direction as it is pinned in the stud. There appears to be no real advantage in this, as breguet springs, unlike flat ones, may be pinned in anywhere, simply at haphazard, and act just as truly and nicely as at equal turns. Watches so sprung by the writer have many times obtained 80 marks and over at Kew, and taken high positions in Greenwich Admiralty trials.

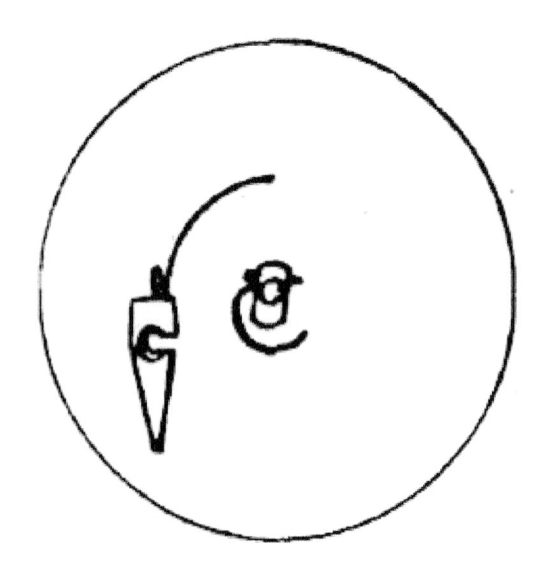

FIG. 160.—BREGUET SPRING PINNED IN AT EQUAL TURNS.

Time the watch by altering the weight of the screws. Additional weight can sometimes be given by changing a pair of brass screws for a pair of gold ones, by changing common gold for good gold, or gold for platinum.

If the watch in hand is a good one, with a compensation balance, it may be desired to adjust it for temperature and positions. At all events, one thing is quite certain; if the watch ever had been so adjusted, by the very fact of re-springing it all adjustment is gone. A new hairspring will require a fresh adjustment for temperature to suit itself, and disturbing the balance screws altogether upsets the position rates.

Good watches often require re-springing. Sometimes a little rust appears on the coils. This is fatal, as it eats further and further in, causing the watch to lose at a gradually increasing rate. Sometimes they get damaged by a careless workman or wearer. From one cause and another a repairer often is called upon to re-spring a good watch, and there is no reason why he should not re-adjust it and turn it out a credit to himself, if he will take the trouble.

Adjusting for Temperature.—When this is to be done, some sort of oven is necessary to keep the watch at about 80° to 90°. The cold can generally be managed in England. In the winter 40° to 50° is easily obtained in a workshop, too easily in fact; while our summers are not so sultry that a few days cannot be found when 55° or 60° can be got without waiting long. With care these temperatures will serve. The oven may be a tin box set on an iron plate, and warmed underneath by being placed on brackets against a wall, about a foot, or less, over a small gas-jet. Or there are many ways that suggest themselves and are suitable to the special conditions of the workshop. A thermometer should be in the oven with the watch.

The watch, when running on time for about two days in the cold, is set by the shop regulator, and its rate noted in a rate book. It is then put in the oven for 24 hours and a comparison made. If 20 seconds slow in heat, it is under-compensated, and one or more screws should be moved several holes nearer to the free ends of the segments. If fast in heat, it is over-compensated, and screws must be moved back. Proceed thus until the watch shows an equal rate, even when tried three days in succession in cold and in the oven. It will be found that moving a pair of screws one hole will make a difference of about 2 seconds per day.

Timing in Positions.—When correct, poise the balance, and again bring to time within about 20 seconds per day. The next thing to do is to get the long and short arcs equal. When a watch is lying down, the balance spins on one pivot and makes a large vibration, say 1 1/2 turns. When it is placed vertically, with 12, 9, or 3 up, the balance runs on the sides of two pivots, and there is more friction. It therefore makes, say, only 1 1/4 turns.

These are, therefore, the long and short arcs, and in all probability the short arcs are slower than the long ones. This is a fault of the hairspring, and the overcoil can be so shaped that the spring will cause the long and short arcs to be made in equal times.

To test the watch, note its rate lying for 24 hours (do not be tempted to make trials shorter than this, as they are misleading). Then 9 up and 3 up for 24 hours each. These two opposite quarters are tried so as to eliminate the errors caused by want of perfect poise. Suppose the rates shown are as follows:—

Watch lying + 3 secs. per day.

Watch 9 up + 11 secs. per day.

Watch 3 up – 5 secs. per day.

Here the rate 9 up is 8 seconds faster than lying, and 3 up 8 seconds slower. The mean of the two is equal to the lying rate, showing that the long and short arcs are equal, and the difference between the two quarter positions is only a question of poising the balance by a touch of a quarter screw.

Suppose the rate is as under:—

Watch lying + 3 secs. per day.

Watch 9 up – 2 secs. per day.

Watch 3 up – 13 secs. per day.

This is a more likely rate, and shows that 9 up the watch is 5 seconds slow, and 3 up 16 seconds slow. The mean is 10 1/2 seconds, which is the amount the short arcs are slow.

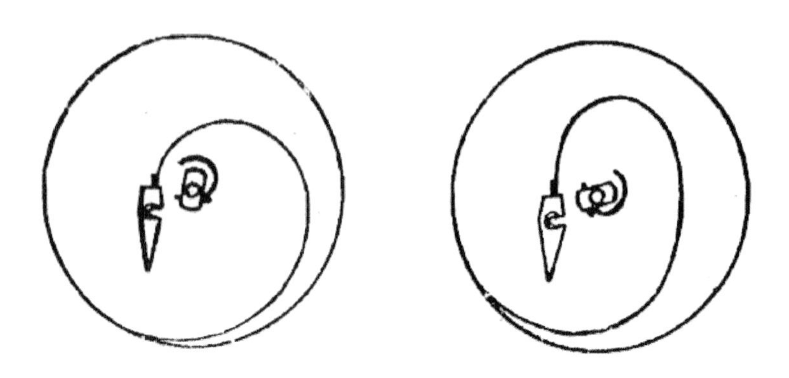

FIG. 161.—FORMS OF OVERCOIL.

To correct this, the form of the overcoil is slightly altered to make it more flexible. Therefore bend the overcoil to make the curve more symmetrical, an easy flow from start to finish, with no angles or straights, is the most flexible, and will quicken the short arcs. Fig. 161 shows the overcoils of two watches that have taken Kew "A" certificates with over 80 marks, and may serve as a guide. If this does not do it sufficiently, cause the spring to open or develop more in the direction opposite to that in which it is pinned in its collet. This can be easily done by a trifling bend, quite unnoticeable, in the overcoil.

It is at this point that there is sometimes a slight advantage in having the spring pinned in at equal turns. When the overcoil is a short one, it is not so easy to cause the spring to develop in the required direction unless it is so pinned in. With a longer overcoil or a double overcoil, this difficulty disappears.

When the mean of the short arcs equals the lying rate, correct the positions by trying 12 up, 9 up, and 3 up. A loss in any position shows that the upper part of the balance in that position is heavy. Put that quarter screw in a trifle, or draw the opposite one out. In this way the position rates can be got equal all round, unless there are escapement faults which cause errors of their own.

Such faults are, an imperfectly poised lever, worn or bent pivots, unequal endshakes of scape wheel and pallets, oval jewel holes, wide pivot holes, unequal banking shake of the lever, or sloping bent banking pins and guard pin.

Also a breguet hairspring should have no play between the curb pins. It should touch each pin, but not be nipped between them or strained by them. Watches with no index, "free sprung," time much better than those with curb pins. One troublesome source of errors is absent in them; but a free-sprung watch that is not perfectly adjusted for temperature and in positions is a nuisance.

The method of position timing just explained is condemned by some writers as bad. They say that alterations of the quarter screws upset the poise of the balance. This is not so. The poising tool at its best is imperfect There is a small error always present. Also the balance as poised on the tool has not the hairspring and collet on. This disturbs the poise in itself, seeing that a part of the weight of the hairspring is borne by the stud and a part by the balance, and there is no means of ascertaining how much or in what manner. The balance is poised as nearly as possible and then put in the watch. There its poising is perfected by the only possible means, viz. noting its rates in the various positions and touching the quarter screws accordingly. It has been recommended that the collet be put on the balance when poising, with its pin and a small piece of hairspring inserted as a refinement. This is absurd, as when subsequently sprung the poise is altogether upset many times by altering the balance screws to bring it to time and adjust for temperature.

A theoretical balance is one having *all* its weight in its rim no friction at its pivots, and not connected in any way with an escapement. For such a balance, a mathematician can plan an overcoil for the hairspring that will make the long and short arcs equal. A number of these curves have been planned and published, and many watchmakers have copied them in the belief that they solve the difficulty. But as the theoretical balance is impossible in practice, so the theoretical curve is inapplicable. Just as friction at the pivots varies, balances have long and short staffs, and escapements differ, so the curve to obtain isochronism differs in each watch, and must be made to suit each individual case by trial.

Variations between the positions of dial up and dial down are caused by some parts of the escapement having unequal endshakes, unequal sized pivots, bent banking pins or guard pin, or it may be a little dirt in a jewel hole.

Fig. 156, at A and B, and Fig. 160 show overcoils made for an index in which the first part of the curve is a portion of a circle. Fig. 161 shows two overcoils for watches with no index—"free sprung," as they are termed. In these no portion of the curve is necessarily circular.

If a very exact temperature adjustment is wanted, the watch should be first adjusted in heat and cold until within three or four seconds per day; then the isochronism—equal times of long and short arcs—got correct. It may then be run several months to settle. All balances go off their temperature adjustment a little as the two metals composing the rim settle to one another. It is then readjusted for temperature, the balance finally poised, and the positions got right last of all.

It is a help to a new balance to settle quickly to warm a brass plate and lay the balance on it, to close up the rim; then to cool it by laying on very cold steel. Repeat the process half a dozen times.

Non-magnetic springs are soft and heavy. They are not so nice to handle, being liable to damage and to being bent out of shape. Their weight makes them shake about in the watch during pocket wear and foul the balance arms, the cock, the curb pins, or the stud. This affects the timekeeping. But they are very necessary in electricians' watches. Non-magnetic balances also are soft and easily bent out of truth.

They often go out of truth with the mere lapse of time, and altogether upset the rate of the watch. Like the springs, they are a necessary evil.

Some Swiss watches have latterly been made with balances and springs of an alloy of nickel steel, that is claimed to be hardly affected at all by changes in temperature; but for exact compensation they cannot yet touch the steel spring and compensation balance.

Very cheap watches with plain balances and springs of this alloy have been made, and in their class are a great advance on the same watch with a plain balance of brass and a steel spring. One use to which this alloy has been put in watch-work is to make hairsprings that require very little compensating. It is found that an ordinary balance compensates too much for them if cut as usual; but if cut at the centres of the rim, leaving four short bi-metallic arms, a stronger and stiffer balance is produced, that at the same time compensates quite enough to correct all the errors. This pattern of balance is found now on many Swiss levers.

"Invar" at one time seemed likely to effect a revolution in both watches and clocks. Pendulum rods for clocks made of it have been tested at Kew, and found to expand only 1/20 in. per mile per degree Centigrade, which is, of course, quite an inappreciable quantity; but in watches the applications of invar are still in an experimental stage, and that they will be equally far-reaching in the near future seems doubtful.

Compensation balances made of invar and brass instead of steel and brass have not proved so satisfactory as was anticipated. Possibly the difference between the expansion of the two metals is too great to be stable.

Watches in which the balance comes at or about the centre of length of the balance staff have been noticed to time better and go more steadily than those in which the balance is at one end of the staff. The reason of this is not very apparent, but it is none the less true. A peculiarity the writer has found in Karrusel watches (in which the balance comes in the centre of the staff and the staff is extremely short) is that, when first sprung and tested for isochronism, the short arcs are found *fast*, whereas in most watches they are invariably more or less *slow* to begin with. An explanation of this is also wanting.

A good watch wanting a breguet spring is worth fitting with one fire hardened and tempered, although these are rather expensive, and cost from 1*s.* 6*d.* to 3*s.* 6*d.* each. For ordinary fair quality work, the usual hard drawn springs, costing about 4*d*, are good enough.

Repairing Hairsprings.—Geneva watches are often badly treated by their wearers. If they stop from any cause, ladies often stir them up with a pin; the inevitable result being a bent or tangled up hairspring. To get such a spring fairly straight again is a task requiring some skill and patience. Take such a spring off the balance and lay it on a glass. Begin at the centre and follow it round, coil by coil, until the first bend or departure from truth is detected. With tweezers and needle-point correct this. Follow round further and correct, until the end is reached. Then proceed to get it flat. Do this in the same way. Begin at the centre; hold the spring up edgeways to the light, and note where it first departs from the flat. With two pairs of tweezers, correct it as in Fig. 157, and proceed to the next point, and so on. Generally speaking, springs that look all tangled up and done for will, on close inspection, be found to have one or two sharp bends, which, when corrected, bring the spring right again as if by magic. Finally, place the collet on a turning arbor and set the spring flat in the turns; affix the stud to the cock or plate, and set it flat in the watch (without the balance), as in Figs. 153 and 159, and see that it passes between the curb pins properly, and that the collet is central with the balance holes.

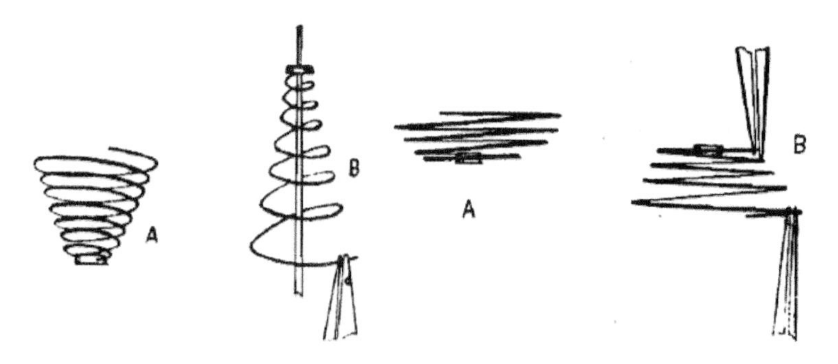

FIGS. 162 AND 163.—FLATTENING THE COILS OF BENT HAIRSPRINGS.

A spring that has been pulled up like A, Fig. 162, may be flattened by putting the collet on a broach and pulling the stud down with tweezers in the reverse direction, as at B. One that is partly pulled up, as at A, Fig. 163, may be treated with two pairs of tweezers, as at B, and sprung flat.

Occasionally in brushing a balance the hairspring gets tangled up, it may be round the balance rim or round its stud. If round the balance rim, lever up the collet and get it off; then turn the spring round and round until it screws off the balance rim. If round its own stud, unpin from the stud, and with a needle-point begin at the centre coil and run the needle round the spiral until the outer end is reached. This will disentangle it safely.

In many watches a shake during wear will frequently jolt the hairspring out of the curb pins, or shake the second coil over or into them. To avoid this, the curb pins should be as long as possible; or in some cases they may be bridged over by arching them at the points and bending towards each other till they touch, making a complete loop enclosing the outer coil. To replace curb pins, file the old ones off flush with the under side of the index, and lay it over a hole in a steel stake, punch them through with a needle, the point of which has been flatted on an oilstone. New curb pins should be filed up nearly straight and well burnished.

In common watches the hairspring usually has some play between the curb pins. This is a source of error, and makes the watch go slower in the hanging position than when lying down. A good watch should be allowed no play here. Each curb pin should just touch the outer coil of the spring lightly without nipping it between them.

Many old watches had plain uncompensated balances and "compensation curbs." A compensation curb was a bi-metallic strip of metal, made so that its movements caused by heat and cold opened and closed the curb pins, thus keeping the watch to time. This was the earliest application of the bi-metallic principle to the compensation of watches, and has been completely superseded by the compensation balance.

From the foregoing, it will be readily understood that an "adjusted" watch takes a long time to get perfect; and its good performance depends upon the fine adjustments of the balance spring and the poise of the balance. When such a watch has an accident and requires a new balance staff, its position rates are at once gone. Several weeks' careful rating are required to re-adjust the poise of the balance.

Careless handling of the balance or hairspring will destroy all the adjustment the watch ever had. Such watches should therefore always receive the greatest care.

THE POWER UNIT

The barrel arbor below the centre of the winding ratchet wheel is always the most sound and sturdy part of the watch. It is a solid little chunk of steel, turned out of a bar with one square and many diameters and a stout hook fashioned on a snail-shaped centre. Its job is to turn with the ratchet wheel and hold the inner end of the mainspring. The spring is fitted to the hook which, in a well-designed watch, does not project from the girth of the arbor, but is cut out of it, so that when tightly coiled the centre wrappings of the spring are snugly circular. A projecting hook causes a kink in the spring which is not good for it.

If the fingers are the power house of the watch the mainspring is its storage battery. A good mainspring will be made of a ribbon of tempered steel, somewhere between seven and fourteen inches in length and of a width and thickness accurately calculated to give the result required by the watch it is to drive. Out of the barrel and left free to choose its own position it will take the shape of a large S with a curly tail. The old mainsprings were complete coils; the newer types are made with a distinct backward turn for the outside end.

Regulating the power of the spring has been one of the watch designer's greatest problems. Beginning with the "stackfreed" some three hundred years ago, then with the fusee and, later, the Geneva stop-work, the provision of a long supple spring of the finest quality steel and the elimination of all unnecessary friction from the movement have proved the best and most economical solution of the problem.

MAINSPRING AS THE REPAIRER RECEIVES IT.

The old stackfreed watches were made almost entirely by hand. Even the gear teeth were hand cut. These watches kept time correct anywhere between thirty minutes and one hour a day. As any better standard was rare, this was accurate enough. Naturally the springs were jerky and uneven, being no more that tempered iron, having the roughest of workmanship in the train and an escapement that was a joke for efficiency. Having a bad timekeeper the old men made it worse with the stackfreed mechanism; it was nothing less than a brake on the spring with a cam arranged in the hope that when the power of the spring was strong it would be restrained, and when weak it would be augmented. The escapement used at that time was the verge, mostly without a balance spring or with a straight wire spring, or even a pig's bristle!

The fusee was a mechanical affair which was highly successful when properly made. It is used to-day on expensive clocks and on ships chronometers. It is an advantage used in connection with a long heavy spring, but of no value in a modern watch. The theory is to give the spring a short lever to pull on when it is at its strongest and long lever when it is weak. The length of the lever varies progressively with the strength of the spring.

In spite or because of hundreds of years of experience of devices to induce the watch spring to release its power evenly and smoothly—including schemes to use two springs in the same watch—the manufacturer now relies on one thin steel ribbon comfortably coiled in a nicely designed barrel to drive his watch. Incidentally it may be mentioned that this type was known as a "going-barrel" watch in its early days, to distinguish it from its then very aristocratic fusee driven relative.

The choice of the correct size and quality of mainspring is a matter of great importance to the timekeeping of the watch. It must be exact in height and thickness to have the required strength, as well as to fit in the barrel without slackness or binding. It must be properly attached at each end. The centre will have a hole to fit over the arbor hook, the outer end should have a short piece of spring rivetted to form a return of which the free end will butt against a recess inside the barrel wall. Fitted like this there is less likelihood of the end breaking off. When springs had the outer coil fastened to the barrel with a hole on a hook, the hook perhaps being stamped out of the barrel wall, there were frequent breakages of the spring or the hook. The rivetted return-end spring is a great advance if only because it lessens the chance of overwinding.

Books could be written on the fragility and waywardness of watch mainsprings. Why they break, apart from sheer old age, is always a mystery and the factory laboratories have almost given up research on the subject. Compared with the number of watches in use and the number of working hours put in by mainsprings failures are very infrequent.

Apart from incorrect fitting, such as inserting a wrong size spring, putting the spring in without lubrication or bad handling when fitting, dampness in the atmosphere is a spring's worst enemy. Rapid changes of temperature cause humidity and small traces of condensation on a spring cause rust. All springs are very susceptible to rust because they work under tension, being incessantly wound and unwound. The action is equal to bending to and fro, which very few materials except indiarubber will stand for long. Spring breakages cannot be explained and cannot be avoided. A poor spring may work for a lifetime; the finest spring may break the day after it is fitted. So far watch mainspring breakage is beyond the skill of man to forecast or prevent. No guarantee can be given in relation to the lasting powers of a mainspring.

Printed in Great Britain
by Amazon